Foundations of Engineering Mechanics

Series Editors: V.I. Babitsky, J. Wittenburg

A.S. Ginevsky · A.I. Zhelannikov

Vortex Wakes of Aircrafts

With 162 Figures

Springer

Series Editors:

V.I. Babitsky
University Loughborough
Department of Mechanical Engineering
Loughborough LE11 3TU, Leicestershire
United Kingdom

J. Wittenburg
Universität Karlsruhe
Fakultät Maschinenbau
Institut für Technische Mechanik
Kaiserstrasse 12
76128 Karlsruhe
Germany

Authors:

A.S. Ginevsky
Central Aerohydrodynamics
Institute (TsAGI)
Radio St. 17
Moskva
Russia 107005

A.I. Zhelannikov
Central Aerohydrodynamics
Institute (TsAGI)
Radio St. 17
Moskva
Russia 107005

ISSN 1612-1384 e-ISSN 1860-6237
ISBN 978-3-642-24246-5 e-ISBN 978-3-642-01760-5
DOI 10.1007/978-3-642-01760-5
Springer Dordrecht Heidelberg London New York

Cover design: deblik, Berlin

Printed on acid-free paper

Springer is part of Springer Science+Business Media (www.springer.com)

CONTENTS

Abstract

The present monograph covers the methods for mathematical modeling of vortex wakes behind aircraft at altitude and near the ground during takeoff and landing operations. The modeling is based on extensive use of the discrete vortex method and on the combination of this method with the integral method for computation of the turbulent boundary layer generated in the near-wall transverse flow induced by an aircraft's vortex system at takeoff and landing. In the latter case, the interaction between the aircraft's vortex wake and secondary vortices caused by boundary-flow separations is taken into consideration.

The methods for simulation of the near and far vortex wakes of turbojet and propeller-driven aircraft are discussed separately. Examples of computed vortex wakes behind some Russian-built and Western aircraft are presented with due consideration for atmospheric turbulence, stratification and crosswinds at landing and takeoff.

In modeling vortex wakes with the proposed methods the required computer execution time is three-four orders of magnitude less in comparison with known numerical methods.

A nonlinear unsteady mathematical model is presented for predicting the motion of light aircraft encountering the vortex wake of a heavy aircraft and a possible severe upset of wake-penetrating aircraft. Computed data are compared with experimental results.

The book is addressed primarily to scientists and engineers of aeronautics but it is also intended for teaching staff, students and postgraduates at aviation institutes and universities.

The present work is an English translation of the book *Ginevsky A.S., Zhelannikov A.I.* Vortex wakes of aircraft. FIZMATLIT, Moscow, 2008, 172 pp., ISBN 978-5-9221-1019-8, published in the original Russian. The publication of this monograph in Russia was supported by the Russian Foundation for Basic Research, project №. 08-01-07081.

Foreword

Investigation of vortex wakes behind various aircraft, especially behind wide-bodied and heavy cargo ones, is of both scientific and practical interest. The vortex wakes shed from the wing's trailing edge are long-lived and attenuate only at distances of 10–12 km behind the wake-generating aircraft. The encounter of other aircraft with the vortex wake of a heavy aircraft is open to catastrophic hazards. For example, air refueling is a dangerous operation partly due to the possibility of the receiver aircraft's encountering the trailing wake of the tanker aircraft.

It is very important to know the behavior of vortex wakes of aircraft during their takeoff and landing operations when the wakes can propagate over the airport's ground surface and be a serious hazard to other departing or arriving aircraft. This knowledge can help in enhancing safety of aircraft's movements in the terminal areas of congested airports where the threat of vortex encounters limits passenger throughput.

Theoretical investigations of aircraft vortex wakes are being intensively performed in the major aviation nations. Used for this purpose are various methods for mathematical modeling of turbulent flows: direct numerical simulation based on the Navier–Stokes equations, large eddy simulation using the Navier–Stokes equations in combination with subrigid scale modeling, simulation based on the Reynolds equations closed with a differential turbulence model. These approaches are widely used in works of Russian and other countries' scientists.

It should be emphasized that the experiments in wind tunnels and studies of natural vortex wakes behind heavy and light aircraft in flight experiments are equally important.

Prof. S.M. Belotserkovsky, a Russian pioneer of theoretical studies of aircraft vortex wakes, showed that the vortex wake problems can be successfully treated with the discrete vortex method developed by this distinguished scientist.

The present monograph brought to the attention of the reader is devoted to further advancements of the ideas related to vortex wake simulation in the works by Belotserkovsky's disciples and followers, published in the Proceedings of the Zhukovsky Air Force Engineering Academy and the Zhukovsky Central Aerohydrodynamics Institute (TsAGI). The discrete vortex method has proven to be the simplest numerical research tool, which requires a significantly smaller amount of computer time when compared to the aforementioned approaches. Besides, the discrete vortex method is used not only for computing aircraft aerodynamic characteristics and studying the origination of trailing vortices, but also for predicting the development of the vortex wakes behind aircraft and other objects (aircraft carriers, buildings, mountains, hills, etc.).

The discrete vortex method has proven to be very efficient for studying steady-state and unsteady flows of an ideal fluid when compressibility can be ignored, and for a closed-form description of free turbulent flows (with $Re \to \infty$) in jets, wakes and mixing layers.

When solving a number of problems the authors use additional empirical information, and in predicting vortex wakes' evolution in the vicinity of the ground they take into account the interaction of the wakes with a wake-induced transverse atmospheric surface flow generating the turbulent boundary layer. It is precisely the interaction of the separating boundary layer with the vortex wake allows one to predict the so-called vortex rebound, when a vortex wake can reach a height of 20–50 m above the runway surface.

It is my belief that this monograph is a serious contribution to the study of this important and complex aviation-related problem.

Academician O.M. Belotserkovsky

Introduction

The monograph brought to the attention of the reader is devoted to numerical simulation of vortex wakes behind aircraft.

Nowadays, the aviation specialists of many developed countries are facing an urgent problem: how to ensure in the future the required passenger throughput at congested airports taking into account the projected air traffic volume for 2015 equal to 2,5–3 times the present figures and to simultaneously decrease the accidental rate of civil aircraft fleet at least by a factor of three. One of the challenges associated with attaining these objectives lies in providing adequate flight safety in the airspace of congested airports. The essence of this peculiar safety problem is in the fact that each flying aircraft generates in the atmosphere behind itself a long-lived vortex wake posing a hazard to other aircraft encountering the wake. The extent of such a vortex wake behind wide-body aircraft is about 10 to 12 km, sometimes even 15 km, depending on atmospheric conditions. At longer distances the wake disappears. This is associated with its natural dissipation and other phenomena. Due to the effect of water vapor condensation, the vortex wake sometimes becomes visible for a ground observer.

The vortex wake depends on the aircraft's design, gross mass, configuration corresponding to a flight phase, flight altitude and speed. Under the action of natural forces the vortex wake sinks 50–300 m below the aircraft's flight path, and also drift horizontally due to a wind and ground effect. Behind an aircraft flying at altitude, the far vortex wake represents two sinking counter-rotating parallel plait-shaped vortex tubes. A decrease in circulation of each of them with time is caused by mutual penetration (diffusion) of vorticities of opposite sign. For an aircraft flying in a turbulent atmosphere, a stronger turbulence intensifies the vorticity diffusion outside the vortex tubes, which results in an additional circulation loss in each of the tubes. At present, there are known different empirical formulas for estimating circulation losses at high and low levels of atmospheric turbulence.

There is also the problem of the interaction of aircraft vortex wakes with the aerodrome's surface during takeoff and landing operations. The problem's importance grows due to continuously increasing congestion of airports. Many countries in the European Union, the USA and Russia, as well as China and India, express concern over this problem. Taking into account the interaction of the vortex wake with the ground surface within an inviscid approximation leads to the well-known result: the vortical structure of an aircraft (two counter-rotating vortices near the ground and their two mirror images relative to the ground surface, forming a quadrupole) is unsteady with both vortices sinking and moving apart. Already initial experimental studies of vortex systems in wind tunnels with a ground board had shown that the wing-tip vortices not only sink and move apart (in accordance with the inviscid approximation), but the so-called rebound of both vortices takes place to a certain height with a subsequent loop-shaped movement of their tracks in transverse planes. Based on experiments, it was shown that such a behavior of the vortices is caused by separation of the boundary layer formed on the ground board as a result of transverse flow induced by the wing's vortical system on the board's surface. Secondary vortices shed into the flow during boundary layer separation interact with the primary ones, resulting in the aforementioned transverse movement of the primary vortices and their rebound.

Today, there are ICAO rules in force, which determine the minimum separation distances between aircraft flying in the same direction (longitudinal separation), which exclude aircraft encounters with vortex wakes. According to these rules, the minimum separations are determined by aircraft types. All aircraft are arbitrarily divided into three classes: light (with weight up to 7 t), medium (between 7 and 136 t) and heavy (greater than 136 t). Thus, the minimum separations are specified, which are, for example, 4 nm (7,4 km) for heavy aircraft following a heavy aircraft, and 6 nm (11 km) for a light aircraft behind a heavy aircraft (Fig. 1,4 in Ref. [1]). With the advent of ultra-heavy aircraft of the A–380 type, the need has arisen for increasing the safe separations between aircraft. According to ICAO recommendations (ICAO Report "Wake Vortex aspects of the Airbus A380 aircraft"11/10/2005: T 13/3-05-0661.SLG), for aircraft following a A-380, longitudinal separations should be equal to the present separations for the corresponding aircraft following a heavy aircraft plus 2 nm (3,7 km) if the follower is a heavy aircraft, or plus 4 nm (7,4 km) if the follower is a medium or light one.

The ICAO rules also specify vertical en route separation of aircraft. The need for increasing air route capacity has already lead to the introduction of six additional flight levels (RVSM program) and the introduction of the minimum vertical separation of 1,000 ft (300 m) instead of the usual 2,000 ft (610 m).

At landing and takeoff on the same runway or on two closely located parallel runways, the permissible time interval between arrivals or departures of aircraft is 2–3 min. However in practice during takeoff and landing operations the vortex wake behind an aircraft often moves away from the runway under the action of external conditions and does not affect the movement of other aircraft. In this case already after 20–30 sec another aircraft can safely land on the runway or can be cleared for takeoff. Under other atmospheric conditions, the vortex can remain over the runway for a long time, posing a hazard for other aircraft. For example, at a crosswind of 1–2 m/s the vortex wake can stay over the runway for several minutes.

Specialists of many countries are trying to coordinate their efforts for creating a special wake vortex avoidance system. These efforts include scientific and engineering conferences on such systems. One of the recent conferences was held in February 2007 in Brussels. Having recognized that further enhancement of air traffic control and flight safety depends on a resolution of this issue, the ICAO has worked out requirements for the future vortex wake-warning system. These requirements are presented in the Air Traffic Control Manual (ICAO Doc 9426, Part II, Chapter 3, Appendix A). The vortex wake-warning systems must consist of ground and airborne subsystems. The fixed minimum separation distances due to vortex wake hazard must be replaced by minimum separations dependent on concrete atmospheric conditions and concrete leader-follower pairs. Besides, these systems must detect dangerous vortex wake zones and not lead to addition workload for air traffic controllers and flight crews.

These requirements are most completely met by the Russian wake-safety management system based on the CNS/ATM ICAO technologies [9, 19, 20]. These technologies are the most promising means for ensuring the effectiveness of air traffic management systems, and, in accordance with an ICAO global plan, must be introduced worldwide by 2010–2020 as a mandatory equipment component of air traffic servicing.

For modeling and studying aircraft vortex wakes a wide range of different theoretical methods are used: direct numerical simulation (DNS) of turbulent flow based on the Navier-Stokes equations, large eddy simulation (LES) using the Navier-Stokes equations and subgrid-scale modeling, as well as numerically solving the Reynolds-averaged Navier-Stokes equations (RANS) with a differential closure model [71, 72]. In the works by S.M. Belotserkovsky it was proposed to use the discrete vortex method [7, 30] for numerically simulating aircraft vortex wakes [8].

The most informative methods from those named above are the DNS and LES methods, which permit studies of near and far vortex wakes at high and low heights from the ground. They allow one in particular to investigate the effects of atmospheric turbulence, atmospheric stratification and windshear as well as the interaction between the aircraft vortex wake and engine exhaust plumes. RANS methods are effective in solving example problems of the interaction of the two oppositely rotating vortex tubes with the ground board. This allows modeling the effects of interaction between the far vortex wake and the ground surface.

Experimental studies of aircraft vortex wakes are performed on models in wind tunnels or in flight tests using laser (lidar) measurement techniques [84].

In recent years three fundamental monographs were published on simulation of aircraft vortex wakes [8, 71, 72]. Refs. [71, 72] are based on the use of different numerical methods for solving a wide range of problems, with Ref. [72] also being devoted to windtunnel and flight experiments for the same purpose. The mathematical modeling methods described in these works allow one to solve the entire range of problems concerning the evolution of the aircraft vortex wakes when flying at height or near the ground during takeoff and landing. Using numerical simulation methods, the authors have given the answers to a number of fundamental questions.

The third monograph [8] is devoted to the creation vortex-wake mathematical model using the discrete vortex method. This method is the most simple and effective tool for investigating the aircraft vortex wake and, being used in combination with some empirical formulas, allows obtaining simple solutions to corresponding problems both for aircraft flying high and near the ground at landing and takeoff.

The discrete vortex method when used for predicting separated flow about bodies can be successfully combined with the methods of unsteady laminar and turbulent boundary layer theory [26]. During takeoff and landing operations, the aircraft vortex system induces in the vicinity of the aerodrome's surface a crossflow accompanied by turbulent boundary layer generation. The vortices formed due to separation of this boundary layer interact with the aircraft vortex system, which results in significant deformation of the latter.

An important feature of the discrete vortex method, as applied to the simulation of the aircraft vortex wake, is that it is equally suitable for prediction of the aircraft vortex wake at every stage of its evolution.

Besides, an important advantage of the mathematical models on the basis of the discrete method is their ease of the use and high computational efficiency. This fact has attracted researchers from many countries to the use of the discrete vortex method for simulation of the far vortex wake and studying its characteristics [73, 76, 77].

The key feature of this monograph is that it is the first to present investigations of the vortex wakes of propeller-driven aircraft and the approaches proposed in the book require the computer execution time on three-four orders of magnitude less as compared with methods based on numerically solving the Navier-Stokes equations. The monograph presents further advancements of the methods described in Ref. [8] and gives their extension for treating a number of new problems.

The present monograph consists of the Introduction and eight chapters.

Chapter 1 presents the basic information about atmospheric turbulence, vortex wakes behind aircraft and modern numerical methods for computing vortex wakes' characteristics.

Chapter 2 describes the discrete vortex method and its usage for simulation of free turbulence in separated and jet flows.

Chapter 3 demonstrates the results of numerical simulation of far vortex wakes behind some aircraft.

Chapter 4 considers a mathematical model of the far vortex wake and presents the characteristics of vortex wakes behind Il-76, An-124, B-747 and A-380 jet transport aircraft.

Chapter 5 presents a mathematical model of the far vortex wake behind propeller-driven aircraft and vortex wake characteristics of An-26, An-12 and C-130 aircraft.

Chapter 6 deals with a mathematical model for computing wind flows over a terrain and presents computed data on air flow over mountains and ravines.

Chapter 7 presents a mathematical model of the vortex wake for aircraft on takeoff and landing and computed data for the vortex wakes behind B-727, Tu-204 and Il-96 aircraft.

Chapter 8 contains a description of a mathematical model for predicting the behavior of airplanes in vortex wakes. Presented in this chapter are the movement of a Yak-40 aircraft in vortex wake environments generated by Il-76, An-124, B-747 and A-380 aircraft as well as the behavior of a Su-25 fighter aircraft in the vortex wake created by a terrain relief and the dynamics of a MiG-31 fighter during its air refueling from a Il-78 tanker aircraft.

The authors wish to thank their colleagues and disciples B.S. Kritsky, S.I. Nekrakha, S.M. Eremenko, S.A. Ushakov, A.V. Golovnev, A.S. Dzuba, N.N. Kopylov for permission to use their material in this monograph.

The monograph was translated into English by Dr. Yu.Ya. Shilov and Dr. I.Yu. Shilov.

Chapter 1

GENERAL

1.1. Atmospheric turbulence

Turbulence is highly irregular flows of a medium with intensive mixing and chaotic variation of parameters. This is one of the most complicated natural phenomena, and its study often requires a philosophically deep insight into the heart of the things. According to an apocryphal story, the famous scientist Theodore von Kármán was quoted as saying that having appeared before the Creator the first thing he would ask him would be to reveal the mistery of turbulence.

The atmosphere is turbulent by its very nature. Usually the atmosphere contains weak or moderate turbulence, more rarely–strong turbulence with characteristic wave lengths longer than 200 m and vertical gusts exceeding 15 m/s.

There are two types of flows of fluid and gaseous media: regular (smooth) and irregular–with significantly mixing medium's volumes and chaotic variations of velocities and other parameters. The former flows are called laminar, for the latter flows the English physicist W.R. Thomson proposed the term «turbulent». Most flows in the nature are of the second, not fully understood type. For description of such flows, probabilistic (associated with averaging in time and space) methods are used. This is due to the fact that it is impossible to monitor fluctuations at each point of the flow.

The flows corresponding to very high Reynolds numbers ($\mathrm{Re} = U_0 b/\nu$) are of prime practical interest. This dimensionless quantity, Reynolds number, includes a "reference"velocity U_0 (exhaust velocity for a jet, flight speed for an aircraft), a characteristic length b (a nozzle diameter or wing chord) and the viscosity of the medium. Reynolds number represents the relationship between inertia forces and friction (viscosity) forces. The typical values of the Reynolds number in aviation are within the range of $\mathrm{Re} = 10^5$–10^7.

In the last few decades, significant progress has been made in studying the fundamental problems of turbulence, mostly due to A. N. Kolmogorov and A. N. Obukhov, their disciples and followers, as well as their forerunners, L. F. Richardson and J. R. Taylor.

At high Reynolds numbers turbulence is generally sought of as an hierarchy of vortices of different sizes, when there are velocity fluctuations

A.S. Ginevsky, A.I. Zhelannikov, *Vortex Wakes of Aircrafts*, Foundations of Engineering Mechanics,
DOI 10.1007/978-3-642-01760-5_1,

in the flow from large to the smallest ones. The large-scale turbulence is determined by the shape of a flow-immersed body and the state of the environment. Here, viscosity forces in formation of vortex wakes can be ignored. When describing small-scale turbulent flows, in some cases a mechanism of molecular viscosity should be taken into consideration.

According to the Kolmogorov–Obukhov theory, the local structure of the developed low-scale turbulence to a large measure obeys the universal laws. It has been proved that in the region of sufficiently small scales a statistical universal regime, practically steady and homogeneous, must dominate.

There is also an intermediate subrange of turbulence — the inertial subrange — corresponding to scales small compared with the characteristic size of the entire flow, but greater than the micro-scale, at which viscosity effects are already significant. Thus, in this subrange, as well as in the initial phase of turbulence, the viscosity of the medium may be not taken into account.

However, the general theory of turbulence, which would provide not only qualitative description of the basic processes, but also quantitative relations allowing obtaining turbulence characteristics, has not been created yet. Construction of a mathematically rigorous theory is hampered by the fact that it is hardly possible to exhaustively define turbulence itself. On the other side, the questions of different engineering applications require immediate answers, even approximate but scientifically substantiated. As a result, the so-called semi-empirical theory of turbulence started its intensive development, where along with theoretical relationships experimental data were used. Such scientists as J.R. Taylor, L. Prandtl and Th. von Kármán have made their contribution to the development of such an approach.

The semi-empirical theory treats the problem in a simplified manner since not all statistical characteristics are addressed, but only the practically most important: mean velocities and means of the squares and cross products of velocity fluctuations (moments of the first and second orders) in the first place. A drawback of such an approach is in the need for experimental data for each set of concrete conditions: for bodies of different shapes when studying vortex wakes, for different configurations of exhaust nozzles, etc. Besides, this theory is based on steady-state approaches (the evolution of a process in time is not considered), which narrows its capabilities.

The works by S.M. Belotserkovsky and A.S. Ginevsky [23, 24] further develop the computer-based vortex concept of turbulent wakes and jets. This concept represents a closed constructive mathematical model based on all achievements of the vortex aerodynamics obtained through the discrete vortex method (DVM). The mathematical model has been constructed for high Reynolds numbers and treats free turbulence as a hierarchy of vortices of different scales. In so doing, turbulent motion is considered in the general case as three-dimensional and unsteady.

In actual practice, modeling unsteady jet flows is achieved by the DVM. For this purpose the model continuous in space and time is replaced with its discrete analogue. Discretization in time consists in the transformation of the continuous process into a stepped analogue with changes at time instants $t_n = n\,\Delta t$ $(n = 1, 2, \ldots)$. Spatial discretization means the replacement of continuous vortex layers with hydrodynamically closed systems of vortex elements (vortex filaments or frames). Besides, it is important that the model takes into consideration the fact that the free vortices move at velocities of fluid particles, whose number increases with time.

The described approach to simulation of flows makes it possible, without invoking additional empirical information, to study the general nature of the time evolution of the process. Mathematical models developed on the basis of the DVM describe all main features of the evolution of turbulent wakes, jets and separated flows, including the transition from deterministic behavior to chaos. They also make it possible to evaluate statistical characteristics of turbulence (moments of the first and second orders). The main emphasis is placed on the computations of flow past bodies and the construction of the portions of vortex wakes immediately behind them. A large amount of data has been accumulated, providing not only direct comparison of computations with experiments, but also verification of the mathematical model for compliance with the universal laws of the Kolmogorov–Obukhov developed turbulence, which in this case act as independent tests. Numerical experiment in combination with physical tests and a comprehensive analysis of the results makes it possible to draw the following conclusions.

The main features and macroeffects of the separated flows about bodies at high Reynolds numbers, including the near wake with account for known separation locations (sharp edges, kinks, base regions) as well as jets, do not depend on the viscosity of the medium. They are determined by inertial interactions in fluids, gaseous or liquid, which are described by unsteady equations for an ideal medium. It has been shown that in a number of problems viscous separations must be also taken into consideration, especially on the surface of smooth bodies (for example, circular or elliptical cylinders). Because of this, the next step in the development of the given concept was to supplement the unsteady models of an ideal medium with the unsteady boundary layer equations to determine separation locations.

Thus, a change in priorities has been substantiated and performed: instead of viscosity, now in the foreground there are unsteady phenomena.

The fundamental work by N. Ye. Zhukovsky «On bound vortices» was published in 1906. The new age has brought forth new problems and computer-based technologies have enlarged the area of application of new theoretical methods. Zhukovsky's classical ideas now have a second youth, opening new opportunities for the ideal medium theory and vortex methods.

The important point is that in the nature vortex flows and chaos live side by side, becoming sources of turbulence. Rotation of fluid volumes generates instability as well as the appearance and breakdown of regular structures, which gives birth to new vortices and development of chaos.

1.2. Aircraft vortex wake

Let us consider a region of disturbed vortex flow behind aircraft. This region, called the vortex wake, is a result of air-aircraft interation and differs from the outside air space in values of velocities, pressures, temperatures etc. From here on, except as otherwise noted, by aircraft are meant both airplanes and helicopters. A distinction should be made between the notions «trailing wake» and «vortex wake». Trailing wakes form as a result of an attached flow past a body in a viscous medium and caused by the body's drag. Trailing wakes represent regions of a decreased mean speed and a higher level of turbulence. Vortex wakes form from the flow around a body in a viscous medium due to generation of lift and, correspondingly, induced drag. They are accompanied by formation, at a distance behind the body, of a pair of counter-rotating longitudinal vortices.

The vortex wake behind an aircraft is an unbounded flow which moves at a mean speed of the undisturbed flow. The length of a vortex wake is about 10–12 km, sometimes greater, and depends on the atmospheric conditions, the aircraft's aerodynamic layout and high-lift devices' position, gross mass, flight speed and altitude. The vortex wake behind an aircraft is characterized by the field of disturbance velocities W, as well as the shape and spatial location of the tip vortices. Disturbance velocities are usually expressed as the following components: W_x — axial velocities; W_R — radial velocities; W_t — tangential velocities, which, in turn, are divided into the vertical, W_y, and horizontal, W_z, velocities, that is,

$$W_t = \sqrt{W_y^2 + W_z^2}\,.$$

Along the vortex wake, the following zones may be arbitrarily specified (Fig. 1.1):

wake formation zone;
stable wake zone;
unstable wake zone;
wake breakdown zone.

In **the wake formation zone**, the roll-up takes place of the entire aircraft vortex structure, boundary layer and engine exhaust jets into two vortex tubes (see a photo on the book's front cover). The boundary layer shed from the airframe surface does not disturb significantly the airflow and practically dissipates at a distance of 50–150 m behind the aircraft.

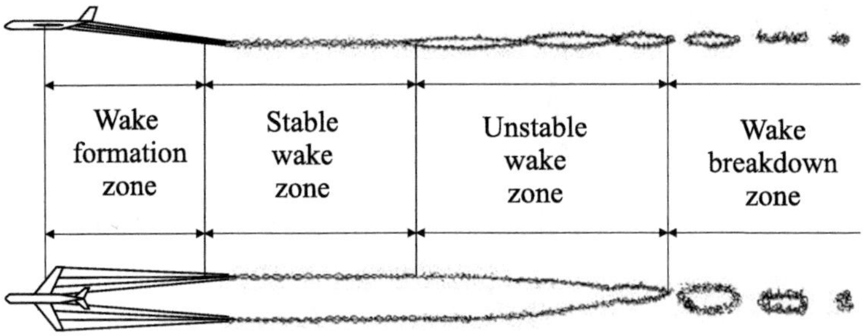

Fig. 1.1. Breakdown of the aircraft vortex wake in accordance with its evolution phases.

The engine exhaust jets with their high kinetic energy represent narrow gaseous flows enlarging at angles of 3–4°; their temperature and velocity quickly decrease.

The boundary layer and exhaust gases influence the initial parameters of the wing-tip vortices: this may manifest itself in increasing the vortices' temperature and velocity.

The greatest contribution to the formation of the trailing wake behind an aircraft and its resulting behavior is made by the vortices shed from the wing, horizontal stabilizer and other lifting and control surfaces, as well as from the fuselage. The vortex sheet just behind the aircraft can be considered in the main approximation as a surface of discontinuity for the tangential velocity component. The thickness of this surface is commensurable with that of the turbulent boundary layer which forms a continuation of the boundary layer shed from the sharp trailing edge and tip edges of the aircraft surfaces. As a result, immediately behind the trailing edge of the wing a turbulent vortex core of finite size is observed. It is a centre around which the wing-tip vortex is formed. By the radius of the vortex core is meant the distance from the vortex tube's axis to the point in its cross section where the value of tangential velocity reaches a maximum. The wake formation zone ends with the formation of steady wing-tip vortices, its length is about 2–3 times the wing span.

In **the stable wake zone** stable movement and sinking of wing-tip vortices take place with their gradual decay. The wing-tip vortices in this zone are stable structures rotating oppositely — the rotation of the left wing's vortex is clockwise, the other vortex rotates counter-clockwise, looking forward. With symmetric loads on both wings, the vortices have equal intensity. Fig. 1.2 demonstrates a typical distribution of vertical velocities in the wing-tip vortices' cores in the stable wake zone of an Il-76 aircraft flying at $H = 400\,\text{m}$ and $V = 550\,\text{km/h}$.

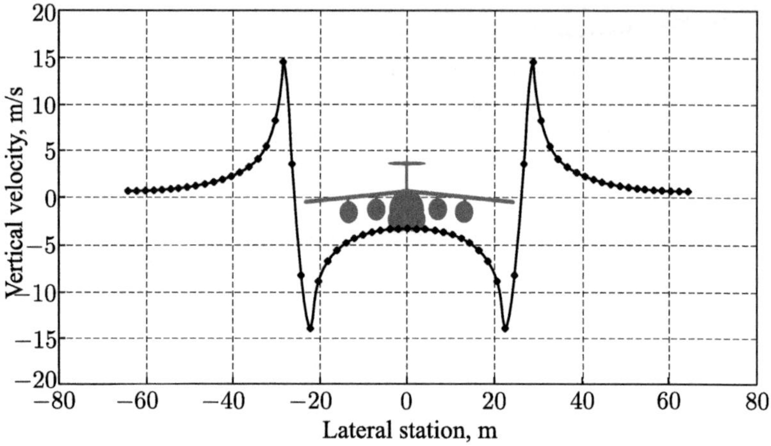

Fig. 1.2. Typical airflow vertical velocity distribution in a control cross-flow plane behind an aircraft

The distance between vortex axes depend on the wing loads. For a symmetrically loaded wing, the distance between the vortices' axes is 0,8L, where L is the aircraft's wing span.

Beginning with the stable wake zone, the trailing wake has a tendency to sinking. The sink of the wake is caused by the vortices' mutual interference and the sink rate is approximately equal to the velocity induced by one wing-tip vortex at the axis of the other. The behaviour of the trailing wake obeys the general laws of vortices' movement in the atmosphere. The velocity field in the trailing wake's zone is significantly nonuniform. The velocities of the disturbed motion of air depend on the intensity of the wing-tip vortices and the wake's age. The flow is characterized by the presence of tangential components of the disturbance velocity and the axial component along the vortex, which can be either equal to the free-stream velocity, exceed it, or be directed against the oncoming flow. The tangential velocity field generates downwash in the zone between the vortices and upwash in the external region (see Fig. 1.2). The maximum values of the circumferential velocities in some cases can reach half the flight speed of the vortex-generating aircraft.

The distribution of axial velocities is alternating in nature. Near the periphery of the core the axial flow is opposite in direction to the flow near the vortex centerline.

The flow regime in the wing-tip vortices is, as a rule, turbulent. The fluctuating velocity components superposed on the averaged velocity field in the wake zone assist in mixing air layers and hastening the decay and demise of the wing-tip vortices. The turbulence is mainly confined to the zone of the vortex core.

An analysis of the results obtained in flight experiments shows that with increasing the length of the trailing wake the entrainment of air into the vortex core' zone, as a rule, weakens.

In **the unsteady wake zone**, the vortices begin to break down. The main modes of the tip vortices' initial breakdown in this zone are as follows:

natural decay due to vortex dissipation and diffusion;
sinuous instability of wake vortices associated with atmospheric turbulence (see a photo on the back cover of this book);
intensive turbulization of the vortex core («vortex burst»), whose mechanism is still scantily known.

Domination of one of these factors determines not only the intensity of the wing-tip vortices' collapse, but also their spatial location relative to the vortex-generating aircraft. In the course of time the structure and position of the wake change. The wing-tip vortex system in the zone under consideration is relatively long-lived, gradually decaying or breaking down.

Sometimes the vortex system executes a symmetric and approximately harmonic motion. The developed trailing wake represents two tip vortex tubes with approximately parallel axes. The vortex lines in each tube have a form of a spiral wound around the tube's axis. In the stable wake zone the axial disturbance velocities directed along the tube's axis and caused by the aircraft's motion influence the spatial location of the vortices and deform vortex lines. The vortex system moves in its own induced flow field. Atmospheric turbulence, wind gusts or random deviations from the flight trajectory deform vortex lines insignifically. Due to mutual induction and the effects of axial velocities the wing-tip vortex tube deforms in a such way that its axis takes an undulating form with the wave length λ and amplitude δ. For certain relations between λ and δ, the wavy character of the tube's axis, and correspondingly of the tube itself, results in fast growing unsteady oscillations. The axes of the vortex tubes come periodically closer and apart until they merge at the nearest points and form isolated vortex rings.

The wake breakdown zone contains a chain of isolated vortex rings. The process of the wake's breakdown is due to sinuous instability and the flow in this zone is unstable. The wake breakdown zone is significantly shorter than the unstable wake zone. Studies show that the vortex wakes behind aircraft of complex aerodynamic configurations, with wings featuring low-aspect and complex planforms, as well as behind helicopters, are prone to sinuous instability to a greater extent than those behind aircraft of conventional configuration.

It should be noted that by wake breakdown is meant a process of changing the trailing wake state which results in restructuring the flow pattern in the wing-tip vortex tubes, damping out disturbance velocities in the tubes and their vicinity to the values commensurable with velocity fluctuations in the atmosphere, changing the spatial location of the wing-tip vortices and the associated alteration of the flow's kinematic parameters.

The vortex breakdown can also occur as a result of the «burst» of the vortex core. By this phenomenon is meant an abrupt expansion of the vortex core and a decrease in the maximum values of circumferential velocities. The physical nature of this phenomenon is supposedly associated with a certain level of axial velocities in the wing-tip vortex, their variation with time and the relation between the axial and tangetial velocities in the vortex tube. The burst can cause the formation of toroidal secondary vortices and reversed flows. The core burst is specific for one of the wing-tip vortices and is a local phenomenon: the other tip vortex can still persist in the atmosphere for a prolonged period.

Variations of the vortex wake's characteristics also take place as a result of interference between vortex flows with buildings, mountains, hills, woodlands. In the aerodrome zone, the formation of local vortex flows is favored by high ground installations, rugged terrain. Although the height of surface constructions near runways is limited, in the aggregate these buildings form a pronounced local obstacle, a barrier for an oncoming flow. The air moves past the upper and side surfaces of these structures, favouring the formation of local vortex flows, including wind shear. The change of vortex wake characteristics becomes more intensive, the boundaries between the unstable wake zones and wake breakdown zones become obliterated, unsteadiness of the wake structure and flow within it becomes more pronounced. The interference of the vortex wake and vortex flows from buildings and rugged terrain creates at the aerodrome complex flow structures, wind shear, and flow velocities dangerous to aircraft.

The presented information on vortex wakes and flows and the trends of their propagation in space allow one to draw a conclusion about unsteady flow in the wake zone. Flow unsteadiness is caused by a temporal relation between velocity variations in a wake cross-sectional plane and spatial wake characteristics. In should be noted that unsteady changes in the location of the wake become more pronounced in the case of sinuous instability.

The above-mentioned peculiarities of the formation and propagation of the vortex wake allow one to consider two approaches to its modeling. The first presents the vortex wake in the form of the wing-tip vortices' axes and the location of the wake is identified with the location of the axes. Disturbance velocities in the wake's zone are specified or computed on the basis of empirical or semi-empirical models, which affects the accuracy of modeling. The other approach models the vortex wake as the evolution of the vortex sheet in the process of simulating the flow past the aircraft under consideration. In this case, the location of the vortex wake is computed directly, whereas the disturbance velocities are determined by the peculiarities of the flow about the aircraft.

1.3. Turbulence characteristics of the vortex wake

As was shown in Section 1.2, the aircraft vortex wake is characterized by the disturbance velocities W, and their components, $W_{y,z}$, as well as fields of disturbance pressures, p.

Among the characteristics of the vortex wake, the most interesting are instantaneous velocities $W_{y,z}$ at a given point $(\overline{y}, \overline{z})$ and a specified time instant τ.

Of practical interest are averaged velocities defined by the integral:

$$\overline{W}_{y,z}(\overline{y}, \overline{z}) = \frac{1}{\tau_n - \tau_0} \int_{\tau_0}^{\tau_n} W_{y,z}(\overline{y}, \overline{z}, \tau)\, d\tau, \tag{1.1}$$

where τ_0, τ_n — instants of nondimensional time corresponding to the beginning and end of averaging, respectively.

Perturbation components of velocity are presented by the following formula:

$$W'_{y,z}(\overline{y}, \overline{z}, \tau) = W_{y,z}(\overline{y}, \overline{z}, \tau) - \overline{W}_{y,z}(\overline{y}, \overline{z}), \tag{1.2}$$

whereas their root-mean-square quantities, representing normal Reynolds stresses, are defined by the integral

$$\overline{W'^2_{y,z}}(\overline{y}, \overline{z}) = \frac{1}{\tau_n - \tau_0} \int_{\tau_0}^{\tau_n} W'^2_{y,z}(\overline{y}, \overline{z}, \tau)\, d\tau, \tag{1.3}$$

The turbulence intensity, an important characteristic from the viewpoint of vortical environment, is defined as

$$\overline{\sigma}_{y,z}(\overline{y}, \overline{z}) = \sqrt{\frac{1}{\tau_n - \tau_0} \int_{\tau_0}^{\tau_n} \left[W_{y,z}(\overline{y}, \overline{z}, \tau) - \overline{W}_{y,z}(\overline{y}, \overline{z})\right]^2 d\tau}\,. \tag{1.4}$$

Besides, for further calculation of the effect of these disturbances on an aircraft, it is important to know the averaged values of the flow angle

$$\Delta\varepsilon = \operatorname{arctg} \frac{\overline{W}_y}{V_0} = \frac{1}{\tau_n - \tau_0} \int_{\tau_0}^{\tau_n} \operatorname{arctg} \frac{W_y}{V_0}(\overline{y}, \overline{z}, \tau)\, d\tau, \tag{1.5}$$

and its fluctuations

$$\overline{\sigma}_{\varepsilon}(\overline{y}, \overline{z}) = \sqrt{\frac{1}{\tau_n - \tau_0} \int_{\tau_0}^{\tau_n} \left[\operatorname{arctg} \frac{W_y}{V_0}(\overline{y}, \overline{z}, \tau) - \operatorname{arctg} \frac{\overline{W}_y}{V_0}(\overline{y}, \overline{z})\right]^2 d\tau}\,. \tag{1.6}$$

V_0 in expressions (1.5) and (1.6) is the undisturbed flow velocity.

1.4. Present-day methods for numerical simulation of vortex wakes behind trunk-route aircraft

Turbulent vortex wakes form behind bodies generating lift. This fact differs them from turbulent wakes behind bodies creating no lift but only drag.

For numerical simulation of aircraft vortex wakes different methods are used, the main of them are the following:

1) Direct numerical simulation (DNS) of turbulent flows on the basis of the full Navier–Stokes equations [11, 21, 36, 40, 49].
2) Large eddy simulation (LES) using the Navier-Stokes equations and subgrid scale modeling [37, 54, 61, 64, 66, 72].
3) Numerically solving the Reynolds-averaged Navier–Stokes equations closed through a differential turbulence model, RANS methods [24, 25, 51, 60].

Among the above methods, the most informative are the DNS and LES, through which it is possible to study the behavior of aircraft vortex wakes at altitude and near the ground. In particular, they offer possibilities to study the effect of atmospheric turbulence and stratification, as well as interaction of aircraft vortex wakes with engine exhaust jets.

Admittedly, DNS methods are tedious and time consuming and what is more, they are effective only for low Reynolds numbers and require the use of supercomputers. LES methods are simpler than DNS ones and make it possible to study the interaction between aircraft vortex wakes and engine exhaust plumes [64, 72].

RANS methods are somewhat more economical. They are effective when solving problems on the interaction of two counter-rotating vortices with the ground surface both in the presence and absence of a crosswind, in simulating the interaction of engine exhaust jets with vortex wakes, when computing the swirling rate of the wing-tip vortex [51, 59, 60, 68, 72].

In a number of cases, LES and RANS methods are used for simulation of 3D turbulent motion, for example, in the problem on the degeneration of the aircraft two-vortex wake system with formation of a chain of vortex rings (a result of the so-called sinuous instability) [41, 72], and estimation of the effect of atmospheric turbulence on this process.

Behind a flying aircraft its far vortex wake represents two sinking parallel counter-rotating vortices. A decrease in the circulation of each vortex with time is caused by mutual penetration of vorticities of opposite signs (the so-called loss of circulation). The afore-mentioned methods of mathematical modeling, in principle, make it possible to evaluate the loss of circulation. In this case, a more intensive loss of circulation with the growth of the level of atmospheric turbulence is associated with an increase in vorticity diffusion with turbulence.

When simulating the aircraft vortex wake using DNS, LES and RANS methods, the initial location and circulation of the wake is usually specified to study further streamwise variation of the wake's position.

Admittedly, compared to the aforementioned methods [8, 11, 13], the most simple and effective is the discrete vortex method (DVM) when using for studying aircraft vortex wakes at high Reynolds numbers. The credit for the promoting such investigations is given to S.M. Belotserkovsky [12, 13]. The indicated approach has been further advanced by works of his followers [8, 14, 17, 18, 29, 31]. This approach is based on the widespread use of the DVM for simulation of vortex wakes in combination with data of field experiments. In particular, this method employs empirical data for the loss of circulation and its dependence on the level of atmospheric turbulence [9, 46, 63], as well as the conclusion made in the framework of the RANS method on a weak effect of engine exhaust jets on the structure of vortex wakes.

The DVM is used not only for studying vortex wakes but also for predicting flow about a wing and a complete aircraft, i.e., for description of the process of origination of the vortex wake, its spatial movement and further evolution.

When studying for vortex wakes (primary vortices) in the vicinity of the ground surface in takeoff/landing regimes on the basis of the DVM, it is possible to close the problem's model only with taking into account viscosity effects, i.e., the formation of secondary vortices induced by the primary ones, by using methods of turbulent boundary-layer theory [14, 39, 52], which make it possible to describe the so-called vortex rebound. Data from Ref. [14, 72] illustrate the DVM's effectiveness in studying the evolution of the aircraft vortex wake near the ground at landing: the computation of the vortex wake at a fixed flight altitude took about 2 min of computer execution time on a medium-scale PC, whereas when using the LES method it takes about 1000 hr.

The present monograph generalized the potentialities of the DVM as applied to simulation of the vortex wakes of trunk-route aircraft.

Chapter 2

DISCRETE VORTEX METHOD

This chapter presents the statement of the problem at hand and fundamentals of the discrete vortex method (DVM). The ideas of this method were for the first time thoroughly and in detail described in Ref. [30] by the S.M. Belotserkovsky and M.I. Nisht. In the succeeding years, further developments of this method were implemented by their disciples and followers [3–6, 22, 26–28, 30, 31].

2.1. Problem statement

Considered here is the unsteady flow of an ideal incompressible fluid past an aircraft flying at the speed W_∞ (Fig. 2.1). The motion of the aircraft and deflection of its control surfaces and high-lift devices are preformed in an arbitrary way. The aircraft's surface is considered impermeable. The flow is potential everywhere outside of the aircraft and vortex wakes generated by flow separation from its surface. The vortex wakes represent thin vortex sheets, i.e., surfaces of discontinuity for the tangential component of the velocity. The lines of flow separation are specified.

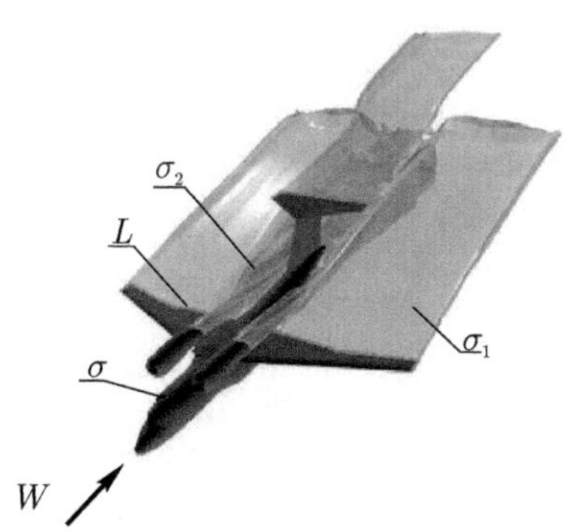

Fig. 2.1. Computational model

Let us denote the lifting and control surfaces of an aircraft together with its engine nacelles by σ, the free vortex sheet shed from the lifting and control surfaces by σ_1, the surface of the exhaust jet by σ_2. The lines of the sheet's shedding are labeled L.

From a mathematical point of view, the problem at hand is reduced to obtaining in a suitable coordinate system the unsteady fields of the velocities $\overrightarrow{W}(\overrightarrow{r}, t)$ and pressures $p(\overrightarrow{r}, t)$, which must satisfy the following conditions and equations:

– The perturbation velocity potential $U(\overrightarrow{r}, t)$ at every time moment outside of the surfaces σ, σ_1 and σ_2 must satisfy the Laplace

A.S. Ginevsky, A.I. Zhelannikov, *Vortex Wakes of Aircrafts*, Foundations of Engineering Mechanics, DOI 10.1007/978-3-642-01760-5_2, © Springer-Verlag Berlin Heidelberg 2009

equation

$$\Delta U = 0. \tag{2.1}$$

— On the surface σ, the flow tangency condition must be met:

$$\frac{\partial U}{\partial n} = -\overrightarrow{W}_{\infty}\overrightarrow{n}. \tag{2.2}$$

— On the vortex wake's surfaces σ_1 and σ_2, being the tangential discontinuity surfaces, the condition of zero pressure jump across the wake at every its point and the no-flow condition through the surface must be satisfied:

$$p^{+} = p^{-}, \quad W_n^{+} = W_n^{-} = V_n, \tag{2.3}$$

where V_n is the normal component of the velocity on the surface σ_1.

— At the separation lines, the Chaplygin–Zhukovsky condition concerning velocity finiteness must be met:

$$\overrightarrow{W}_n(\overrightarrow{r}, t) \to 0. \tag{2.4}$$

— At infinity, the disturbances die away:

$$\Delta U \to 0 \quad \text{at} \quad \overrightarrow{r} \to \infty. \tag{2.5}$$

— For relation between velocity and pressure, the Bernoulli equation is used:

$$p = p_{\infty} + \frac{\rho W_{\infty}^2}{2} - \frac{\rho W^2}{2} - \rho \frac{\partial U}{\partial t}. \tag{2.6}$$

When solving the problem, the potential $U(\overrightarrow{r}, t)$ or $U(M, t)$ is sought in the form of the double-layer potential

$$U(M_0) = \overrightarrow{W}_{\infty}(t) + \sum_{i=1,2} \frac{1}{4\pi} \int_{\sigma_i} \frac{\partial}{\partial \overrightarrow{n}_{\mathrm{M}}} \left(\frac{1}{MM_0}\right) g_i\,(M, t)\, d\sigma_{\mathrm{M}}, \tag{2.7}$$

where $g_i\,(M, t)$ is the density of the double-layer potential on the surface σ. In this case the fluid velocity at every point not lying on the surfaces σ, σ_1 and σ_2 is determined by the formula

$$\overrightarrow{W}(M_0, t) = \overrightarrow{W}_{\infty} + \sum_{i=1,2} \frac{1}{4\pi} \int_{\sigma_i} \nabla_{M_0} \left(\frac{\partial}{\partial \overrightarrow{n}_{\mathrm{M}}} \left(\frac{1}{r_{MM_0}}\right)\right) g_i\,(M, t)\, d\sigma_{\mathrm{M}}. \tag{2.8}$$

Relationship (2.8) is also true on the surfaces σ, σ_1 and σ_2 if the integrals involved in it are meant as hypersingular in the sense of Hadamard's finite value. It will be recalled that the double-layer potential undergoes a jump on surfaces where it is defined, but its normal derivative is continuous. Correspondingly, the velocity field has a jump in the tangential velocity component on the surfaces of the schematized aircraft and its wake, whereas the normal component on these surfaces is continuous.

To satisfy conditions (2.3), we seek such a solution where the surfaces $\sigma_1\,(t)$ and $\sigma_2\,(t)$ consist of points moving together with the fluid, and the density of the double-layer potential, $g_i(M,t)$, at every such point does not depend on time. Suppose that at each instant of time $\tau \leqslant t$ a fluid particle leaves the line of wake-shedding, $M(s)$, where s is the arc length, and at the instant t occupies the position $M(s,\tau,t)$, and at each instant $t$the totality of the points $M(s,\tau,t)$ forms the surface of the vortex wakes $\sigma_1(t)$ and $\sigma_2(t)$. In this case the equation of motion for these surfaces takes the form:

$$\frac{\partial \vec{r}(s,\tau,t)}{\partial t} = \vec{W}(M(s,\tau,t),t), \tau \leqslant t, s: M(s) \in L, \tag{2.9}$$

with the initial conditions

$$\vec{r}(s,\tau,t)\Big|_{t=\tau} = \vec{r}_{M(s)}, \tag{2.10}$$

where $\vec{r}(s,\tau,t)$ and $\vec{r}_{M(s)}$ are the position vectors of the points $M(s,\tau,t)$ and $M(s)$, respectively, whereas for the function $g_2(M,t)$ the following relation is true:

$$g_2(M(s,\tau,t),t) \equiv g_2(s,\tau), \quad \tau \leqslant t, \quad s: M(s) \in L. \tag{2.11}$$

Condition (2.2) is equivalent to the equation

$$\frac{1}{4\pi}\sum_{i=1}^{2}\int\limits_{\sigma_i} \frac{\partial}{\partial \vec{n}_{M_0}} \frac{\partial}{\partial \vec{n}_M}\left(\frac{1}{r_{MM_0}}\right) g_i(M,t)d\sigma_{i,M} = f(M_0), \; M_0 \in \sigma_1, \tag{2.12}$$

where $f(M_0) = -\vec{W}_\infty \vec{n}(M_0)$.

Finally, the interrelation between the functions $g_1(M, t)$ and $g_2(s, t)$ is described by the following formula, resulting from the integrability condition for the velocity field:

$$g_2(s,t) = g_1(M(s), t, s: M(s) \in L). \tag{2.13}$$

Thus, the problem of unsteady separated flow of an ideal fluid past an aircraft is reduced to the solution of the closed system of equations (2.9)–(2.13) for the functions $\bar{r}(s,\tau,t)$, $g_1(M,t)$, $g_2(s,\tau)$. With that, if these functions are the solution of the indicated equations, the potential $U(M,t)$ defined by formula (2.7), the corresponding velocity field $\vec{W}(M,t)$, defined by expression (2.8), and the pressure $p(M,t)$ determined by integral (2.6) satisfy conditions (2.1)–(2.6).

The aircraft's geometry was represented with a combination of thin plates and solid elements. The wing and other lifting surfaces are presented schematically as their surfaces, whereas the fuselage and engine nacelles are modeled with solid elements. The plates and solid elements, in turn, are modeled with a double layer of continuously distributed doublet singularity approximated with the network of discrete closed vortex frames. In this case, rectangular vortex frames (cells) are used. Located along the contour of each cell are vortex filaments, whose intensities are unknown. These vortex filaments induce velocities in accordance with the

Biot–Savart law. The resulting velocity field is sought in the form of the sum of the velocities induced by all vortex frames modeling the body's surface and its wake and the velocity of the oncoming flow:

$$\vec{W}(\vec{r}, t_k) = \sum_{i=1}^{N} \Gamma_i(t_k)\vec{W}_i(\vec{r}) + \sum_{m,l} \Gamma^1_{m,l}\vec{W}_{mlk}(\vec{r}) + \vec{W}_\infty, \qquad (2.14)$$

$$\vec{W}_i(\vec{r}) = \frac{1}{4\pi} \oint_{\vec{r} \in \partial\sigma_i} \frac{[\vec{r} - \vec{r}_0] \times \vec{dl}}{[\vec{r} - \vec{r}_0]^3}.$$

Thereafter the problem is reduced to determining the intensities of the vortex frames representing the body Γ_i and vortex wake $\Gamma_{m,l}$, along with the coordinates of corner points of the vortex frames, $\vec{r}_{m,l}$. For determining intensities Γ_i for each vortex frame, a control point (collocation point) is established in a special way, for which the flow tangency condition is written. Is a result, one obtains the following system of algebraic equations in Γ_i:

$$\sum_{i=1}^{N} \Gamma_i(t_k)\omega_{i,j} = f_j^k, \quad \mathrm{j} = 1 \ldots N, \qquad (2.15)$$

$$\omega_{i,j} = \vec{W}_i(\vec{r}_j)\vec{n}_j, \quad f_j^k = \left[-\sum_{m,l} \Gamma^1_{m,l}\vec{W}_{mlk}(\vec{r}_j) - \vec{W}_\infty \right] \vec{n}_j. \qquad (2.16)$$

When simulating the vortex wake, it is assumed that the vortex frames moves together with fluid particles, and as this takes place, their intensities $\Gamma_{m,l}$ remain constant:

$$\vec{r}_{m,l}(t_k) = \vec{r}_{m,l}(t_{k-1}) + \vec{W}(\vec{r}_{m,l}(t_{k-1}), t_{k-1})\Delta t, \quad l < k. \qquad (2.17)$$

At each instant of time, a new vortex cell forms with its two corner points lying on the separation line:

$$\vec{r}_{kl}(t_k) = \vec{r}^{\,L}_l, \qquad (2.18)$$

and the intensity of the vortex filament of the newly shed frame is determined through the intensities of the vortex filaments lying on the body's surface and having with it a common side:

$$\Gamma_{m,l} = \Gamma_{i^+}(t_m) - \Gamma_{i^-}(t_m). \qquad (2.19)$$

In formulas (2.14)–(2.19), l is the number of the segment of the separation line left by the frame $b_{m,l}$, m is the point in time at which the frame leaves the line.

Thus, the solution of the problem is obtained by time stepping until the specified end of computation. At each step, the loads are computed through the Cauchy–Lagrange integral.

2.2. Fundamentals of the discrete vortex method

The problem in hand is solved by the DVM according to which a flow-immersed body and its wake are replaced with systems of bounded and free vortices (Fig. 2.2). In this case, the closed rectangular vortex frames (cells) are used as hydrodynamic singularities (Fig. 2.3).

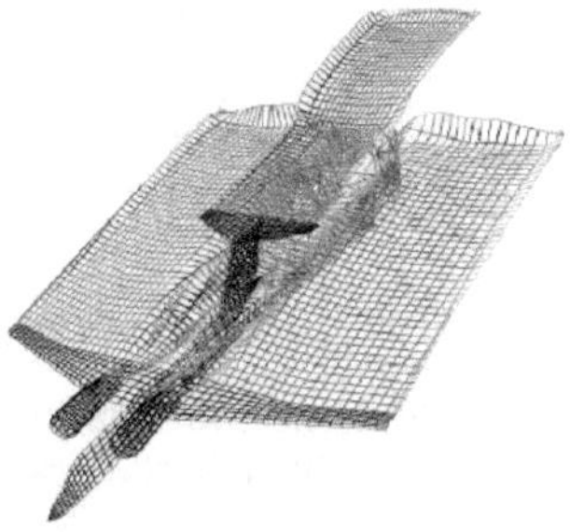

Fig. 2.2. System of vortex frames

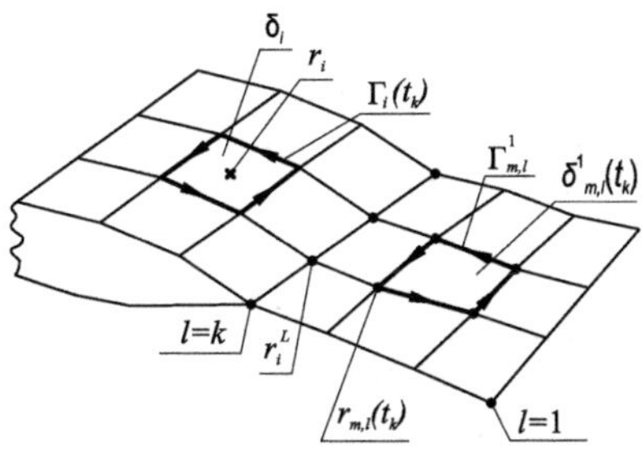

Fig. 2.3. Closed vortex frames

Positioned along the contour of each cell i is a vortex filament whose intensity is unknown. The vortex filaments induce velocities according to the Biot-Savart law. The combined velocity field is sought as the sum of the velocities induced by all vortex frames modeling the body's surface and its wake and the velocity of the oncoming flow:

$$\vec{W}(\vec{r}, t_k) = \sum_{i=1}^{N} \Gamma_i(t_k)\vec{W}_i(\vec{r}) + \sum_{m,l} \Gamma^1_{m,l}\vec{W}_{mlk}(\vec{r}) + \vec{W}_\infty, \qquad (2.20)$$

$$\vec{W}_i(\vec{r}) = \frac{1}{4\pi} \oint_{\vec{r} \in \partial\sigma_i} \frac{[\vec{r} - \vec{r}_0] \times \vec{d}l}{[\vec{r} - \vec{r}_0]^3}.$$

Thereafter the problem is reduced to determining the intensivities of the vortex frames representing the body Γ_i and vortex wake $\Gamma_{m,l}$ along with the coordinates of the corner points of the vortex frames $\vec{r}_{m,l}$. For determining intensities Γ_i for each vortex frame, a control point (collocation point) is specified in a special way, for which the flow tangency condition is written. As a result, one obtains the following system of algebraic equations in Γ_i:

$$\sum_{i=1}^{N} \Gamma_i(t_k)\omega_{ij} = f_j^k, \quad \mathrm{j} = 1, N, \qquad (2.21)$$

$$\omega_{ij} = \vec{W}_i(\vec{r}_j)\vec{n}_j, \quad f_j^k = \left[-\sum_{m,l} \Gamma^1_{m,l}\vec{W}_{mlk}(\vec{r}_j) - \vec{W}_\infty \right] \vec{n}_j. \qquad (2.22)$$

When modeling the vortex wake, it is assumed that the vortex frames moves together with fluid particles and as this takes place their intensities $\Gamma_{m,l}$ remain constant:

$$\vec{r}_{m,l}(t_k) = \vec{r}_{m,l}(t_{k-1}) + \vec{W}(\vec{r}_{m,l}(t_{k-1}), t_{k-1})\Delta t, \quad l < k. \tag{2.23}$$

At each instant of time, a new vortex cell is formed with its two corner points lying on the separation line (2.24):

$$\vec{r}_{kl}(t_k) = \vec{r}_l^{\,L}, \tag{2.24}$$

and the intensity of the vortex filament on the newly shed frame is determined through the intensities of the vortex filaments lying on the body's surface and having with it a common side:

$$\Gamma_{m,l} = \Gamma_{i^+}(t_m) - \Gamma_{i^-}(t_m). \tag{2.25}$$

In formulas (2.20)–(2.25), l is the number of the segment of the separation line left by the frame $b_{m,l}$, m is the point in time at which the frame leaves the line.

Thus, the solution of the problem is obtained by time stepping until the specified final time step is made. At each step, the loads are computed through the Cauchy-Lagrange integral. These loads are averaged over time when needed.

2.3. Point vortex

For simulating a planar flow in aerohydrodynamic problems using the DVM, point vortices, or vortices of infinite length, are used as hydrodynamic singularities. Let us consider a point vortex at point x_0, y_0

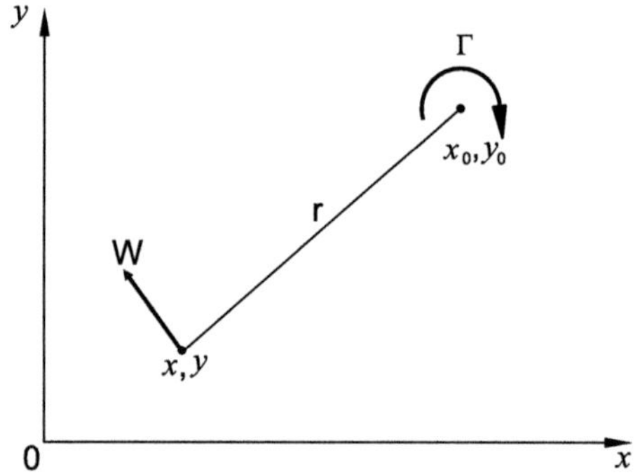

Fig. 2.4. Point vortex

with circulation Γ (Fig. 2.4). The velocity induced by this vortex at each point x, y of the flow plane is determined by the Biot-Savart formula:

$$W = \frac{\Gamma}{2\pi r}, \tag{2.26}$$

where r is the distance from the vortex to an arbitrary point, whereas the components of the velocity W in a suitable coordinate system are computed with the formulas

$$W_x = \frac{\Gamma}{2\pi}\left(\frac{y_0 - y}{r^2}\right), \quad W_y = \frac{\Gamma}{2\pi}\left(\frac{x - x_0}{r^2}\right), \tag{2.27}$$
$$r = \sqrt{(x - x_0)^2 + (y - y_0)^2}.$$

Thus, for predicting vortex wakes in planar flows, calculations of disturbance velocities can be performed using formulas (2.27).

2.4. Vortex segment

The vortex segment can be used as the basic element in computing three-dimensional vortex wakes using the DVM. Consider the velocity field induced by a vortex segment (Fig. 2.5).

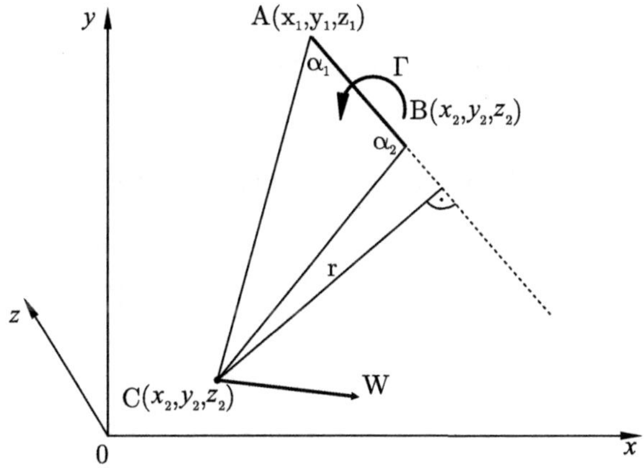

Fig. 2.5. Vortex segment

Let the vortex segment with circulation Γ have the coordinates of its end points $A(x_1, y_1, z_1)$ and $B(x_2, y_2, z_2)$. Then the velocity induced by this vortex at any point $C(x, y, z)$ of the space under consideration will be determined with the Biot–Savart formula

$$W = \frac{\Gamma}{4\pi r}(\cos\alpha_1 + \cos\alpha_2), \tag{2.28}$$

where r is the shortest distance (perpendicular) from any arbitrary point to the vortex segment or its extension (see Fig. 2.5), whereas the components of the velocity W in an adopted coordinate system are computed with the formulas

$$W_x = \frac{\Gamma}{4\pi}ca_x, \quad W_y = \frac{\Gamma}{4\pi}ca_y, \quad W_z = \frac{\Gamma}{4\pi}ca_z, \tag{2.29}$$

where

$$c = \frac{1}{h^2}\left[\frac{(x-x_1)(x_2-x_1)+(y-y_1)(y_2-y_1)+(z-z_1)(z_2-z_1)}{r_1} - \right.$$
$$\left. - \frac{(x-x_2)(x_2-x_1)+(y-y_1)(y_2-y_1)+(z-z_1)(z_2-z_1)}{r_2}\right], \quad (2.30)$$

$$\begin{aligned} r_1 &= \sqrt{(x-x_1)^2+(y-y_1)^2+(z-z_1)^2}, \\ r_2 &= \sqrt{(x-x_2)^2+(y-y_2)^2+(z-z_2)^2}, \\ h &= a_x^2+a_y^2+a_z^2, \\ a_x &= (y-y_1)(z_2-z_1)+(z-z_1)(y_2-y_1), \\ a_y &= (z-z_1)(x_2-x_1)+(x-x_1)(z_2-z_1), \\ a_z &= (x-x_1)(y_2-y_1)+(y-y_1)(x_2-x_1). \end{aligned}$$

Thus, in simulating three-dimensional vortex wakes, formulas (2.29) may be used for computing disturbance velocities.

2.5. Closed vortex frame

The closed vortex frame (Fig. 2.6) can be used as the basic element in computing three-dimensional vortex wakes using the DVM. Usually, such an element is a quadrilateral vortex. For this approach to simulation,

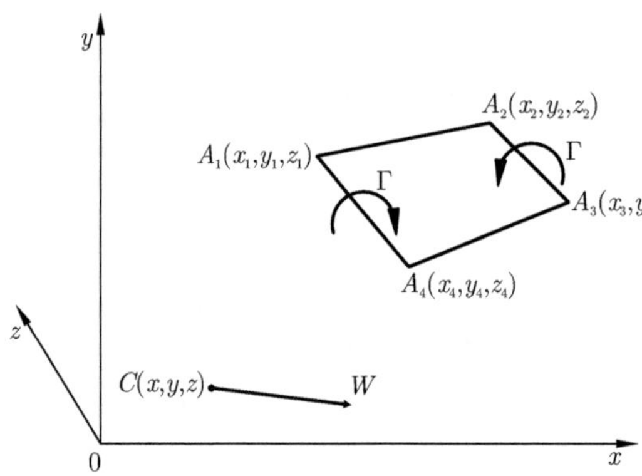

Fig. 2.6. Closed vortex frame

a mathematical base has been created and required mathematical reasons proved, including ones relating to the location of the control point [27]. The velocity induced by the closed vortex frame at any point $C(x, y, z)$

of the space under consideration will be determined as the sum of the velocities induced by the four segments forming the vortex frame:

$$W = \sum_{i=1}^{4} W_i, \tag{2.31}$$

the corresponding components of the velocity W in an adopted coordinate system are computed with formulas

$$W_x = \sum_{i=1}^{4} W_{xi}, \quad W_y = \sum_{i=1}^{4} W_{yi}, \quad W_z = \sum_{i=1}^{4} W_{zi}. \tag{2.32}$$

Thus, in simulating three-dimensional vortex wakes, formulas (2.31) may be used for computing disturbance velocities.

2.6. Numerical modeling of free turbulence in separated and jet flows in the framework of the discrete vortex method

The construction of mathematical models of shear turbulent flow of an incompressible fluid at high Reynolds numbers is based on the treatment of free turbulence as an hierarchy of vortices of different scales. In so doing, the large-scale turbulent motion in the general case is considered as three-dimensional and inherently unsteady; it is originated by loss of stability and breakdown of organized vortex structures and their transformation into vortex ensembles. The latters, moving along with the medium, deform, entrain each other and form both new macrostructures and small vortices.

It is essential that vortex motions contain an inherent mechanism of loss of stability and transition from order to chaos. Solutions obtained through the DVM allow determination – without using empirical constants – of mean velocity and pressure fields, normal and shear Reynolds stresses, pressure fluctuations, correlations between velocity and pressure fluctuations, corresponding turbulence scales and spectra.

The heart of the concept in hand is a discrete description of a phenomenon both is space and time. It can be argued that vortex motions of a liquid medium are governed by an inherent mechanism of loss of stability and transition from order to chaos. The given approach made it possible to confirm an important role of large-scale coherent structures in mixing layers, jets, wakes and separated flow, which were earlier revealed experimentally. Coherent, organized structures — vorticity clots — are localized in space and have significant lifetimes. For the methodology described here, it is very important to establish the fact of a weak medium viscosity dependence of these phenomena, i.e., to a first approximation, viscous dissipation is not taken into account.

Computations of vortical structure of turbulence in wakes, jets, mixing layers and separated flows are performed by solving the system of differential equations

$$\frac{d\overline{x}_i}{d\tau} = \frac{u_i}{u_o}, \quad \frac{d\overline{y}_i}{d\tau} = \frac{v_i}{u_o}, \quad \frac{d\overline{z}_i}{d\tau} = \frac{w_i}{u_o}, \quad i = 1, \ldots, N, \tag{2.33}$$

where i is the number of a free vortex, N is the number of vortices, u_i, v_i, w_i are the velocities at the corners of the vortex frame, induced by all vortices, x_i, y_i, z_i are nondimensional coordinates, τ is the nondimensional time. The complete solution of the problem is obtained in two interdependent phases: solution of a system of linear equations in circulations of all vortices on the body's surface, with account for the flow tangency boundary condition, and determination of free vortices location. Both procedures must be performed simultaneously, however this is usually made with a regular delay of one time step $\Delta\tau$.

All algorithms in one way or another contain implicit disturbance sources, for example, the growth of the number of free vortices during flow development. Because of this, even if the pattern of flow about a body is generally periodic, the initial conditions will not be strictly repetitive at corresponding points in time. Besides, in problems with symmetric conditions, a different sequence of computation of velocities or circulations at symmetric points turns out to be a peculiar kind of disturbance source.

In the right-hand side of equations (2.33) there are nondimensional velocities obtained by summing mean velocities of oncoming flow and their fluctuations and also velocities induced by the vortices on the body and free vortices of a wake or jet. For the vortices the flow tangency boundary conditions are taken into consideration on the surface of the body or nozzle. The main source of turbulence at high Reynolds numbers is the movement of a large number of free discrete vortices. The process of breakdown of regular vortex structures is three-dimensional in character with the leading part playing by inertia forces.

In the DVM there are two mechanisms of energy dissipation (vortex diffusion).

1. As noted above, the movement of free vortices is described by ordinary differential equations (2.33). It is impossible in the DVM to compute velocities in the vicinity of vortices in a «discreteness zone» due to the singularity of the Biot–Savart formula ($u_i = \Gamma_i/2\pi r$); because of this, if a free vortex Γ_i falls in the «discreteness zone» of a vortex Γ_j, it is necessary to «smooth» velocities, which is equivalent to correspondingly decreasing Γ_i, Γ_j. This leads to cutting off velocity peaks, which can be interpreted as «numerical diffusion» in the DVM.

2. In numerically integrating equations (2.33) using the Euler method, at each step $\Delta\tau$ instead of the formula

$$\overline{x}_i^{s+1} = \overline{x}_i^s + \int\limits_{\tau_s}^{\tau_s+\Delta\tau} u_i(\tau_1)d\tau_1, \ldots,$$

its simplified version is used:

$$\overline{x}_i^{s+1} = \overline{x}_i^s + \overline{u}_i^s \Delta\tau, \ldots,$$

i.e., the equations are treated in the following form

$$\frac{d\overline{x}_i}{d\tau} = \overline{u}_i + \varepsilon_x, \quad \frac{d\overline{y}_i}{d\tau} = \overline{v}_i + \varepsilon_y, \quad \frac{d\overline{z}_i}{d\tau} = \overline{w}_i + \varepsilon_z,$$

where ε_x , ε_y, ε_z are small stochastic functions.

As noted above, this methodology provides a closed description of shear turbulence without empirical constants thanks to the fact that in the framework of the DVM it is possible to model dissipation of turbulent energy. In this case there is no need to ensure the coincidence between quantitative dissipation parameters and the true ones. It is important here that the method constructed for simulation of turbulent motion exploits an energy sink.

For computing turbulence statistics, the means values of the velocity components $< u_i >$, mean pressure $< p >$ and their fluctuations $u_i'(\tau)$ and $p'(\tau)$ must be determined:

$$< u_i > = \lim_{T\to\infty} \frac{1}{T} \int_0^T u_i(t+\tau)d\tau, \qquad < p > = \lim_{T\to\infty} \frac{1}{T} \int_0^T p(t+\tau)d\tau, \quad i = 1, 2, 3;$$

$$u_i'(\tau) = u_i - < u_i >, \qquad p'(\tau) = p - < p >,$$

thereafter other characteristics must be obtained: Reynolds normal and shear stresses, correlation coefficients and spectra. The Reynolds normal stresses are determined as follows:

$$< u'^2 > = \lim_{T\to\infty} \frac{1}{T} \int_\tau^{\tau+T} u'^2(\tau),$$

$$< v'^2 > = \lim_{T\to\infty} \frac{1}{T} \int_\tau^{\tau+T} v'^2(\tau)d\tau,$$

$$< w'^2 > = \lim_{T\to\infty} \frac{1}{T} \int_\tau^{\tau+T} w'^2(\tau)d\tau.$$

The Reynolds shear stresses $< u'\, v' >$, $< u'\, w' >$, $< v'\, w' >$ are determined in a similar way.

The correlation coefficients are defined as follows:

The space correlation coefficient for velocity fluctuations

$$R_{uu} = \frac{< u'(x)u'(x+r) >}{\left[< u'^2 > (x)\right]^{1/2} \left[< u'^2 > (x+r)\right]^{1/2}},$$

the correlation coefficient for velocity and pressure fluctuations

$$R_{up} = \frac{< u'(x)p^1(x+r) >}{\left[< u'^2 > (x)\right]^{1/2} \left[< p'^2 > (x+r)\right]^{1/2}},$$

The corresponding integral scale is given by the formula

$$L = \int_0^\infty R_{uu} dr,$$

the time correlation, or autocorrelation, coefficient is defined as

$$R(\tau) = \frac{< u'(\tau) u'(t+\tau) >}{\left[< u'^2 > (\tau)\right]^{1/2}}.$$

The autocorrelation coefficient R (τ) and energy spectrum $E(f)$ are related by the following expressions:

$$R(\tau) = \frac{1}{< u'^2 >} \int_0^\infty E(f) \cos 2\pi f t df,$$

$$E(f) = 4 < u'^2 > \int_0^\infty R(\tau) \cos 2\pi f t dt.$$

If should be noted that unlike the method for computing turbulent jets and wakes using algebraic or differential turbulence models and containing some empirical constants, the computational method described in the given section is free from empirical constants and has one more advantage. It allows determination, apart from mean velocity fields, of three components of Reynolds normal and shear stresses, temporal and spatial-temporal correlation coefficients, velocity fluctuation and pressure correlation coefficient as well as spectra.

To illustrate the capabilities of the above approach to simulation of free shear turbulent flows let us consider the computational results of some applications.

A planar turbulent wake behind a transverse plate placed normally to the flow [23, 25]. Computations show that in this case two flow regimes are possible: symmetric and asymmetric. The first of them turns out to be unstable with no transverse mixing of vortex clots with positive or negative circulation. An asymmetric vortex structure of the planar wake behind the plate at a fixed time instant is shown in Fig. 2.7, which also demonstrates averaged flow patterns behind the plate in asymmetric (*1*) and symmetric (*2*) regimes. In the first case the reverse flow zone is much shorter. Variations of computed and measured values of the mean velocity

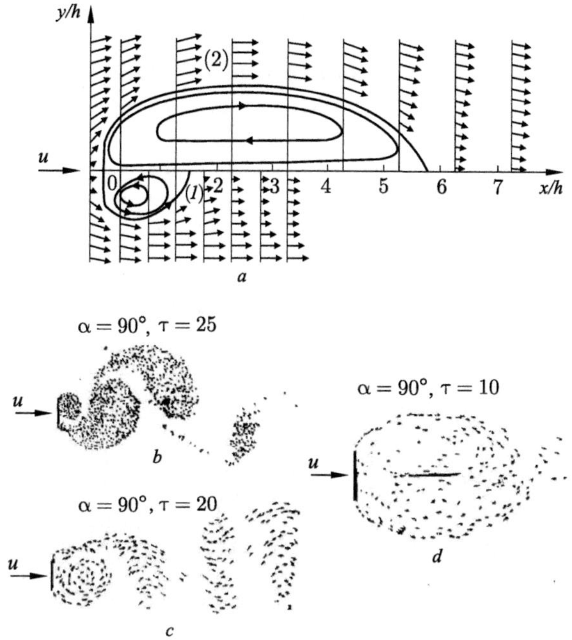

Fig. 2.7. Vortex structure in the wake downstream of a vertically positioned plate. (*a*) Averaged patterns of separated flow in asymmetric (*1*) and symmetric (*2*) regimes; (*b*) — nondimensional time, u — free-stream flow velocity, h — plate chord, — angle of attack.

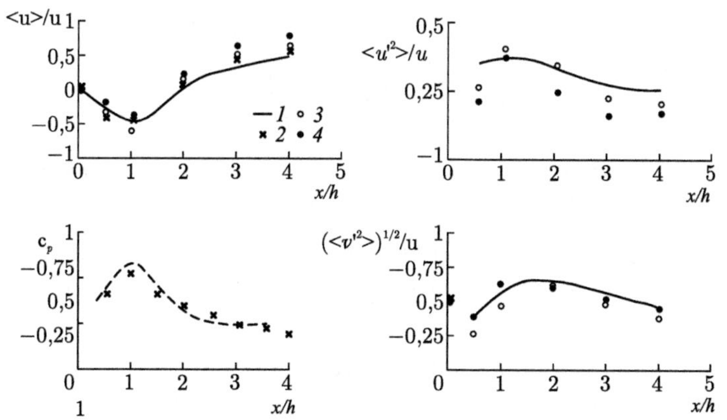

Fig. 2.8. Variation of the mean velocity $< u > /u_\infty$, pressure $c_p =, < p > > /(0{,}5\rho u_\infty^2)$, streamwise $(< u'^2 >)^{1/2}/u_\infty$ and transverse $(< v'^2 >)^{1/2}/u_\infty$ пульсаций velocity fluctuations along the axis of the wake downstream a flat plate over an interval $x/h = 0 - 4$: *1* — experiment; *2–4* — computation [23, 25]

and pressure as well as the two components of velocity fluctuations downstream of the plate are shown in Fig. 2.8.

Separated flow past a spoiler [67, 70]. Compared here are computed [67] and measured [70] velocity profiles $< u >$, velocity fluctuations ε_u, mean pressure $< c_p >$ and the intensities of its fluctuations $\varepsilon_p = (< c_p' >)^{1/2}$ on a plate (Fig. 2.9). Fig. 2.10 demonstrates the streamwise variation of the space correlation coefficient of wall pressure fluctuations $R_{pp}(\dot{x}_0 \Delta x)$ for three values of x_o/h.

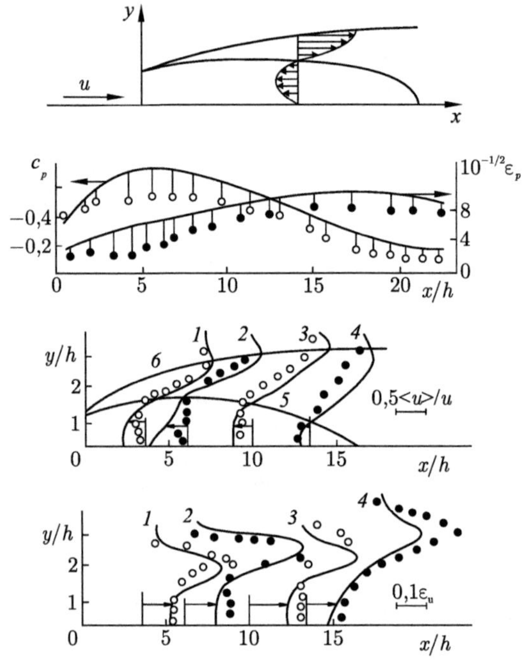

Fig. 2.9. Comparison of computed and experimental data on mean values of pressure $< c_p >$ and pressure fluctuations $\varepsilon_p = (< p'^2 >)^{1/2}/0{,}5\rho u_\infty^2$ on the wall in the separated flow zone behind the spoiler as well as mean velocities profiles $< u > /u_\infty$ and streamwise velocity fluctuation $\varepsilon_u = (< u'^2 >)^{1/2}/u_\infty$: *1–4* — sections, *5* — line of zero streamwise velocities, *6* — circulation zone boundary [67, 70].

Planar immersed turbulent jets [1, 23, 24] and a mixing layer in two semi-infinite flows [55]. As in the case of a planar flow behind a plate, computations show the possibility of occurring two regimes of a planar jet issuing from a nozzle – symmetric and asymmetric. The corresponding vortex structures at a fixed moment of time $\tau = t\ u_o/h = u$ are presented in Fig. 2.11. In actual conditions the asymmetric jet vortex structure occurs, which corresponds to transverse mixing of vorticity clots of both

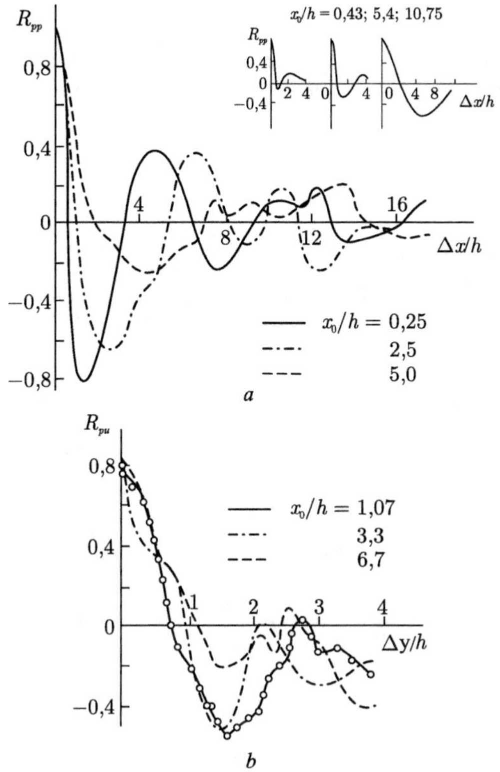

Fig. 2.10. Variation of correlation coefficients for the streamwise wall pressure fluctuations R_{pp} $(x_o,\ \Delta x)$. Experimental data [70] are presented at the top right of the figure

signs. It can be seen from a comparison of computed and measured data for the mean velocity along the jet axis (see Fig. 2.12) that in the case of the asymmetric vortex structure the agreement between theory and experiment is satisfactory.

Circular turbulent jets [23–25]. Used in computations as a basic vortex element was a vortex ring. However, the strict condition of axial symmetry resulted in a very weak enlargement of the jet. Because of this, the problem was treated as three-dimensional, i.e., computations of a circular jet were performed with no requirement for axial symmetry. As basic computational elements, vortex polygons were used, the jet boundaries were modeled with vortex frames, and in the course of extension of the vortex segments of these frames, the segments were divided into smaller ones.

The computation showed (Fig. 2.13) that the nearly circular vortices (vortex polygons) maintain their azimutal uniformity over the first three

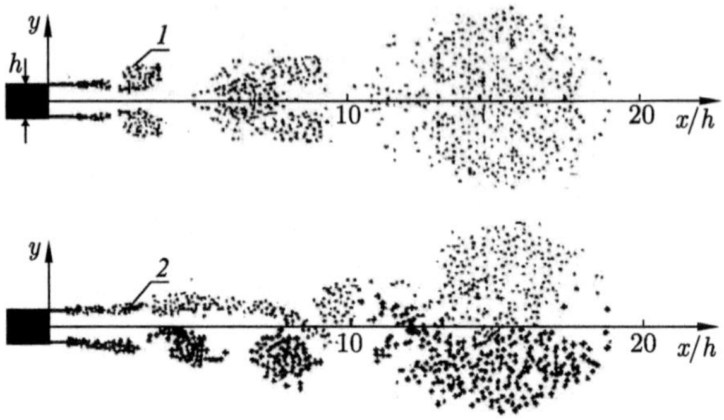

Fig. 2.11. Symmetric (*1*) nd nonsymmetric (*2*) vortex structures in a planar turbulent jet at the moment of tim $\tau = tu_o/h = 45$

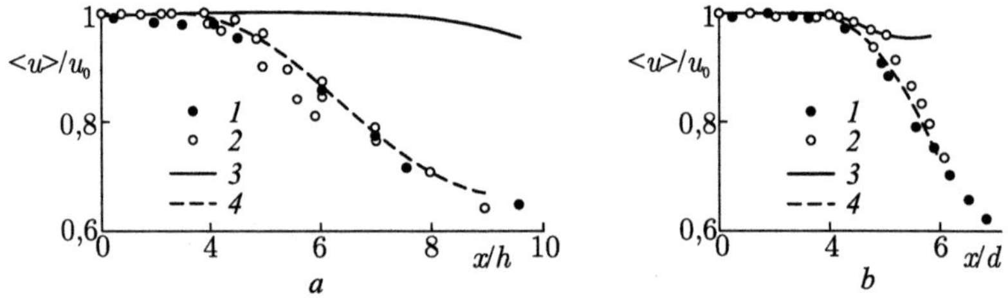

Fig. 2.12. Variation of the velocity along the axis of jets [23, 24]. *(a)* Planar jet: *1* and *2* — experiments, *3* and *4* — computations of flow with symmetric and nonsymmetric vortex structures, respectively; *(b)* Circular turbulent jet: *1* and *2* — experiments *3* and *4* — computations of flow with symmetric and three-dimensional nonsymmetric vortex structures, respectively)

calibers, $x/d = 0$–3, following which within $x/d = 3{,}5$–$6{,}0$ the vortex rings begin to take on a star-shaped and three-dimensional structure; further, at $x/d > 6{,}0$ flow stochastization takes place. With that, as opposed to the axisymmetric approximation, the jet enlargement and the decrease in the mean velocity along the jet axis (Fig. 2.14) are simulated with significantly better agreement between theory and experiment for longitudinal and radial velocity fluctuations and, additionally, azimutal velocity fluctuations are calculated (Fig. 2.14).

The approach presented in this section are used below (see Chapter 6) in calculating a turbulent flow over a terrain. In particular, this approach makes it possible to predict the drag of two- and three- dimensional bodies (plate, prism, disc, etc.).

Results of investigations performed in recent years on a supercomputer into the structure of turbulence in an immersed circular jet [66] are presented in Fig. 2.15. Turbulent flow was considered in a jet of an inviscid fluid and its statistical properties were studied. The velocity field of interacting vortex tubes was obtained on the basis of the Biot–Savart

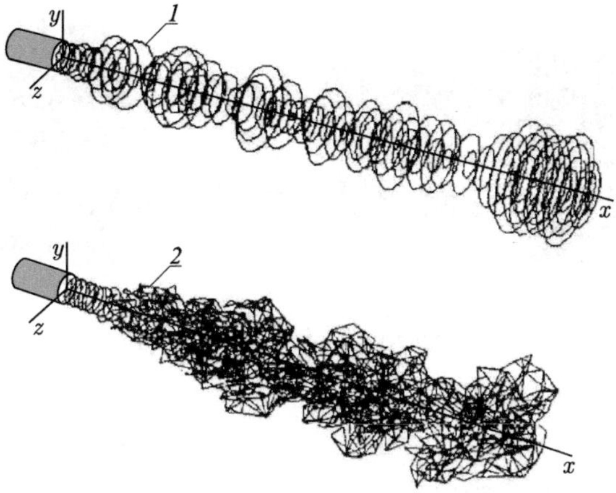

Fig. 2.13. Vortex structures of a circular jet for axisymmetric (*1*) and three-dimensional (*2*) problem formulations at two fixed instants of time $\tau_1 = tu_o/d = 52$ and $\tau_2 = 44{,}4$ [23–25].

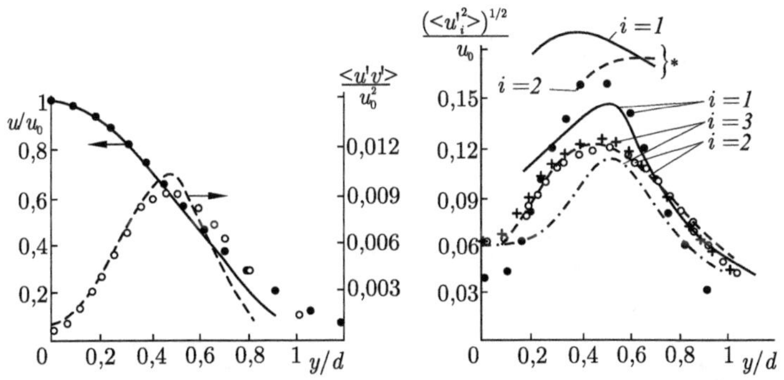

Fig. 2.14. Comparison of computed and experimental profiles of mean velocity $< u > /u_o$, three components of velocity fluctuations $(< u_i' > ^{1/2})/u_o$ $(i = 1, 2, 3)$ and the Reynolds shear stresses $< u'v' > /u_o^2$ in a cross section at $x/d = 4$ computing a jet in a three-dimensional problem formulation (points — experiment; curves - computation; $u_1' = u'$, $u_2' = v'$, $u_3' = w'$) [23–25]

law. It was shown that in the framework of this approach it is possible to obtain turbulence characteristics agreed with physical experiments and data of direct numerical simulation, structure functions, energy spectra, log-normal vorticity distribution and two-point correlation functions.

Swirling turbulent flow in a cylindrical vortex [23]. Let us consider computation of a swirling flow in a cylindrical vortex for the case of an ideal fluid in the framework of the DVM (Fig. 2.16 *a*). It is known that for such a flow there is the exact solution: uniformly distributed vorticity $\omega = \text{const}$ inside the cylinder of radius R.

Fig. 2.15. Coherent vortex structures of the initial region of a circular turbulent jet [66]

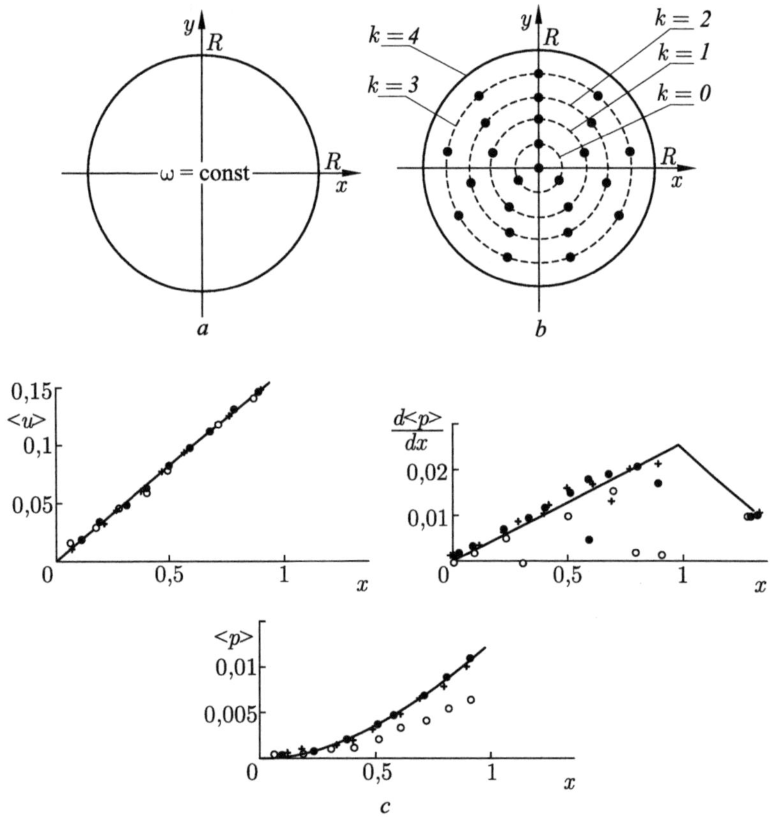

Fig. 2.16. Computational results for an example flow: a — vortex cylinder; b — vortex cylinder discrete approximation; c — comparison of approximate and exact solutions for the circumferential velocity component $< u >$, pressure $< p >$ and pressure gradient $d < p > /dx$: (solid line — exact solution, (•) — $\Delta = 0{,}05$, $N = 400$, (+) — $\Delta = 0{,}1$, $N = 400$, (∘) — $\Delta = 0{,}2$, $N = 400$

In the discrete representation of the vortex cylinder, the vortex interior is filled with discrete vortices with the same circulation Γ_i, uniformly distributed over each circumference (Fig. 2.16 *b*). Nondimensional quantities were defined as follows:

$$\overline{x} = \frac{x}{R}; \quad \overline{y} = \frac{y}{R}; \quad \overline{u} = \frac{uR}{\sum\limits_{i=1}^{N} \Gamma_i}; \quad \overline{p} = \frac{p}{\rho}\left[\left(\sum_{i=1}^{N} \Gamma_i\right) \cdot \frac{1}{R}\right]^{-2}.$$

A comparison of computed velocity pressure and its radial gradient with the exact solution for a cylindrical vortex (Fig. 2.16 *c*) demonstrates satisfactory accuracy of computation with the number of vortices of $N = 400$ and a discreteness measure of $\Delta = 0{,}2$ (inviscid approximation).

The capacity of the DVM has been demonstrated in computing various turbulent flow characteristics, including velocity fluctuations in circumferential and radial directions, turbulence energy and spectral data.

Interaction between a pair of vortices and a flat ground board. In conclusion of this section let us consider a two-dimensional unsteady problem on the interaction between two parallel vortex tubes opposite in sign, propagating over a flat ground board parallel to the latter. [1]) Considered in this case are secondary flows caused by viscosity of the medium and the formation on the ground board a turbulent boundary layer at high Reynolds numbers.

So, the two parallel vortex tubes with circulations Γ_0 and $-\Gamma_0$ are moving parallel to the flat ground board at a distance H_0 from it; the separation distance between the vortices' axes is $2z_0/H_0 = 1$ (Fig. 2.17*a*). To meet the tangency condition on the board, two mirror image vortex tubes with circulations whose signs are opposite to those of the real vortices are placed at a distance of $y = -H_0$ from the ground board surface. The real vortex tubes and their mirror images are represented by a system of 19 rectilinear vortex filaments with equal circulations. With that, the circular core is replaced by the central linear vortex and two concentric layers consisting of 6 and 12 vortices identical with the central one.

The indicated vortex tubes induce a near-wall crossflow accompanied at high Reynolds numbers by the formation of a turbulent boundary layer. The Reynolds number is defined by the formula $\mathrm{Re} = V_0 H_0/\nu$, where V_o is the velocity induced by one of the vortex tubes at the centerline of the other at the initial instant of time. When this layer separates in the region of positive pressure gradient, transverse vortices are generated with circulation of opposite sign. These secondary vortices induce the lateral

[1]) Ginevsky AS, Pogrebnaya TV, Shipilov SD (2009) On the interaction of a vortex pair and vortex ring with a flat ground board (in Russian). Eng.-Phys. Journal (in press)

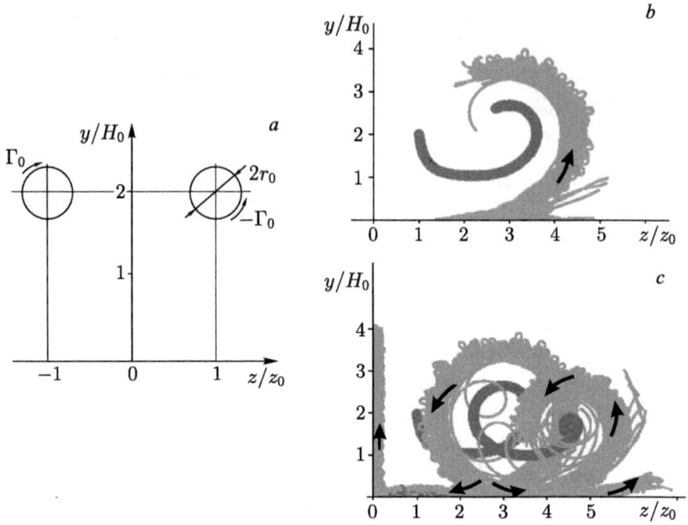

Fig. 2.17. . Interaction between a vortex pair and a flat ground board. Flow pattern at the initial instant of time (*a*). Tracks of the primary (black) and secondary (grey) vortices in a control plane at the instants of time $tV_0/H_0 = 75$ (*b*) and $tV_0/H_0 = 150$ (*c*)

movement of the primary vortices, which results in their loop-shaped tracks in a control plane. This problem was solved within a quasi-steady approximation.

With that, the integral method of boundary-layer calculation is used only for determination of the parameters of secondary vortices produced during the boundary layer separation. In what follows, already in the framework of an ideal medium, the interaction of the primary and secondary vortices is studied. The results of this analysis are shown in Fig. 2.17 *b,c*. Presented here are the traces of the primary and secondary vortices in the control plane at two instants of time and the evolution of the secondary vortices and their action on the lateral movement of the primary vortices and the formation of their loop-shaped traces in the control plane. It is significant that the secondary vortices interact with the primary ones without mixing with them. In other cases, for example, during impingement of the vortex tube at the flat board, the mixing of the secondary and primary vortices takes place.

The problem under discussion is of interest as applied to studying the interaction between the aircraft vortex wake and the aerodrome surface during takeoff and landing operations. The solution of this problem is described in detail in the Chapter 7 of the present book.

Chapter 3

THE NEAR VORTEX WAKE BEHIND A SINGLE AIRCRAFT

3.1. Aircraft geometry representation

The schematic representation of an aircraft's geometry is an important part of modeling the flow about the aircraft and its vortex wake. The present monograph generalizes the approaches of many researchers employing the DVM. Among the types of aircraft representation developed up to now, the following three are the most popular and successful: plate-element, solid-element and hybrid representations.

The plate-element representation idealizes an aircraft as an array of thin plates. Each solid element of the aircraft (fuselage, engine nacelles) is represented by two mutually perpendicular plates («cruciform scheme»). This simplest representation gives good results in many cases, including the prediction of the behavior of an aircraft encountering the wake of another aircraft.

Fig. 3.1. gives by way of example the plate-element representation (developed by S.M. Yeremenko and A.V. Golovnev) for the MiG-29 fighter aircraft.

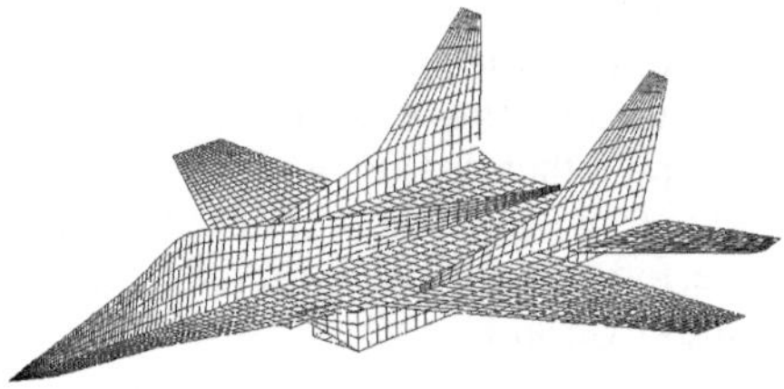

Fig. 3.1. Plate-element representation of the MiG-29 fighter aircraft.

The solid-element representation considers all components of an aircraft (fuselage, engine nacelles, wing, stabilizer, fin) as a combination of solid elements. This is the most complicated representation, but it practically always gives good results, especially in predicting the motion of aircraft.

Presently, the most popular aircraft representation is the hybrid one. In this case the fuselage and engine nacelles are considered as solid elements, whereas the horizontal tail, fin and other lifting and control surfaces

A.S. Ginevsky, A.I. Zhelannikov, *Vortex Wakes of Aircrafts*, Foundations of Engineering Mechanics,
DOI 10.1007/978-3-642-01760-5_3, © Springer-Verlag Berlin Heidelberg 2009

are treated as thin plates. Experience has shown that the best results are obtainable if the element representing the wing is bent according to its mean-camber surface. Fig. 3.2. demonstrated a hybrid representation of the MiG-29 fighter. This hybrid scheme is also developed by S.M. Yeremenko and A.V. Golovnev.

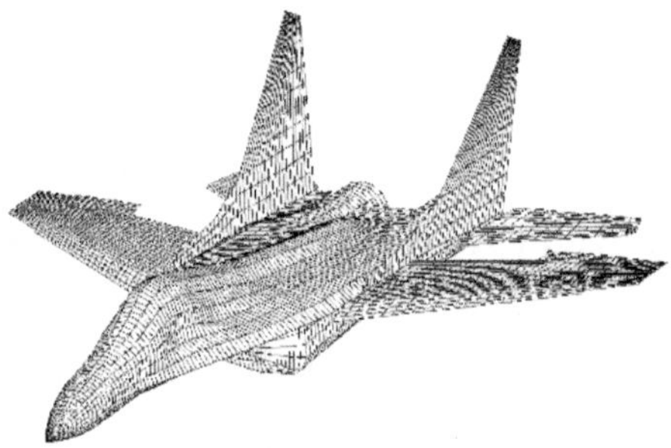

Fig. 3.2. Hybrid geometry representation of the MiG-29

3.2. Vorticity panel representation

Based on a chosen aircraft geometry representation, a network of vorticity panels on the aircraft surface is constructed for the computation of the near vortex wake and the behavior of an aircraft in the wake of another aircraft or in air flow over a terrain. In the present monograph for modeling vortex wakes the hybrid geometry representation of aircraft is used, whereas for predicting the effect of the vortex wake on aircraft dynamics a plate-element representation is preferred. Distributed over the schematized surfaces of the aircraft were vortex frames. Separation lines were fixed; as a rule, these were the sharp edges of the wing, horizontal stabilizer and other lifting and control surfaces. The process of vortex sheets' shedding from these lines was modeled. The discrete time-stepping was performed until the aircraft's aerodynamic loads "level off", i.e., become time invariant.

3.3. Peculiarities of flow simulation around trunk-route aircraft

When simulating flow around trunk-route aircraft and their vortex wakes, some peculiarities must be taken into account. The modern trunk-route aircraft have, as a rule, wide fuselages and thick ($\overline{c} > 8$–10%) wings with effective high-lift devices. Because of this, for obtaining credible computed data on characteristics of the vortex wake it is necessary to use a hybrid aircraft geometry representation and model the fuselage and engine nacelles with solid elements and the wing and other lifting and control surfaces with mean-camber surfaces of these components. As an example, Figs. 3.3–3.5 demonstrate vortex element arrays for computation of vortex wakes behind the A-320, the A-380, and the Il-76 aircraft, respectively. The vorticity panel representation for these aircraft was developed by S.M. Yeremenko.

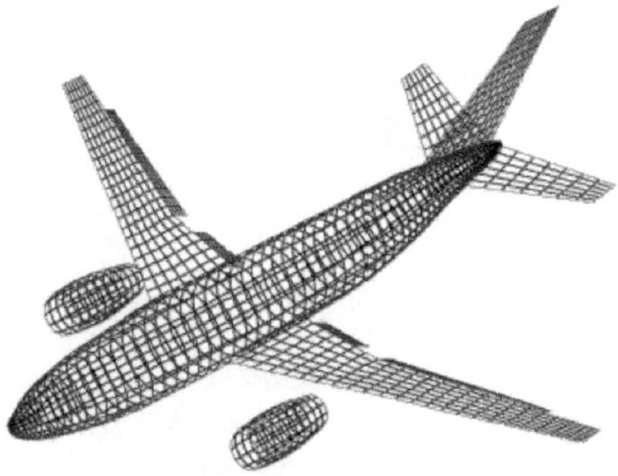

Fig. 3.3. . Vorticity panel representation of the A-320

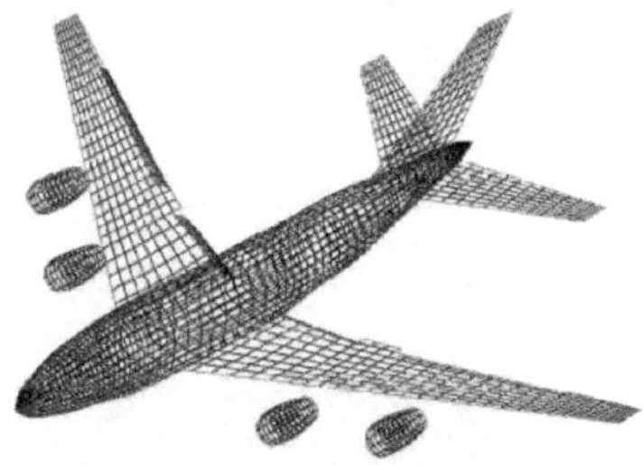

Fig. 3.4. Vorticity panel representation of the A-380

3.4. The characteristics of the near vortex wake behind some aircraft

For predicting flow about aircraft and obtaining the characteristics of their vortex wakes, nonlinear unsteady theory [2, 27, 30] based on the DVM was employed. Some works [2, 27, 28, 30] contain the results of computation of aircraft near wakes with the help of linear and nonlinear steady theories. The computational results for the near vortex wake in the present monograph were obtained using a nonlinear unsteady theory, the most general and universal. Fig. 3.6. demonstrates the computed near vortex wake behind the An-72 aircraft at an angle of attack of $\alpha = 5°$.

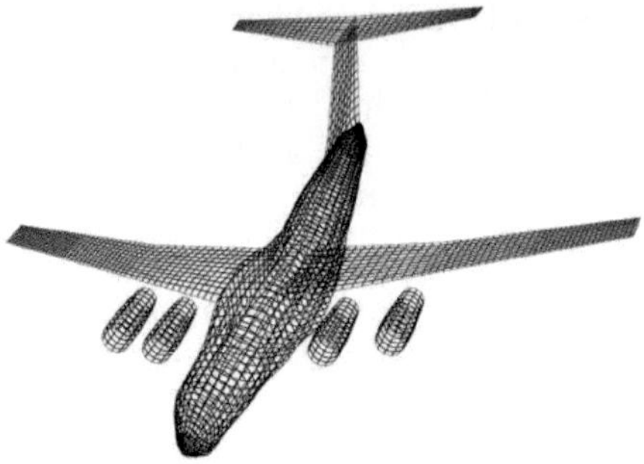

Fig. 3.5. Vorticity panel representation of the Il-76

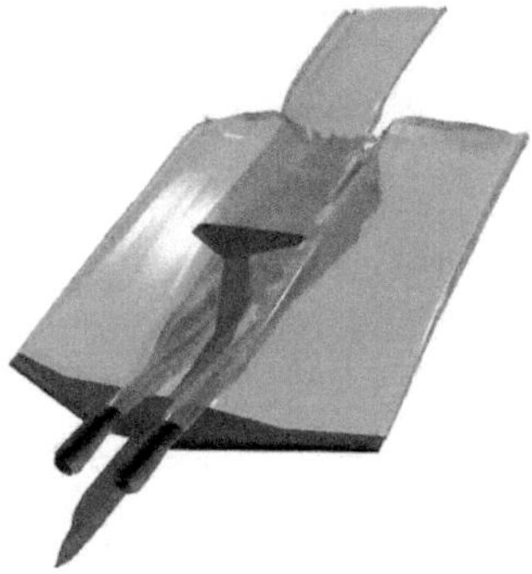

Fig. 3.6. Near vortex wake behind the An-72

Fig. 3.7. Near vortex wake of the Il-76

Its lifting and control surfaces along with the fuselage were modeled with thin plates, whereas the engine nacelles were represented with solid elements.

Fig. 3.7 presents the near vortex wake behind the Il-76 aircraft (at $\alpha = 7°$). Its lifting and control surfaces are modeled with thin plates, and

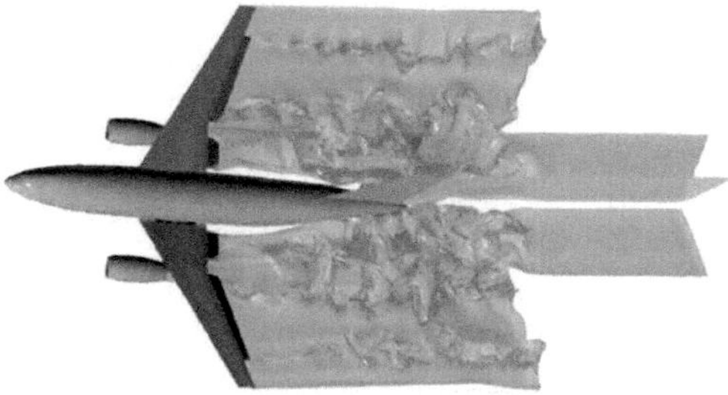

Fig. 3.8. Near vortex wake of the Tu-204

the fuselage and engine nacelles with solid elements. Its high-lift devices of the wing are in landing configuration.

Shown in Fig. 3.8 is the near vortex wave generated by the Tu-204 aircraft at $\alpha = 7°$. Thin plates were used to represent the lifting and control surfaces of the aircraft, solid elements modeled the fuselage and engine nacelles. The high-lift devices are deflected for landing.

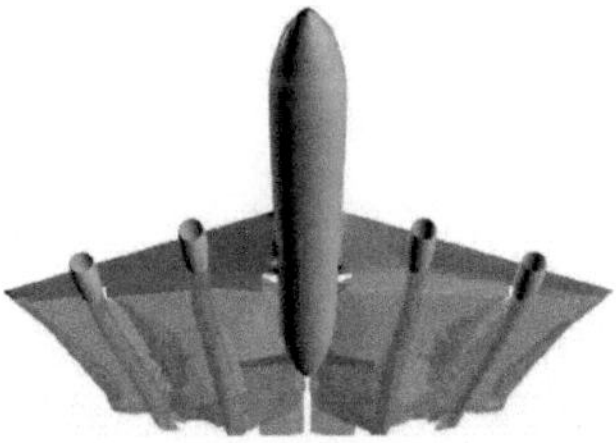

Fig. 3.9. Near vortex wake of the A-340

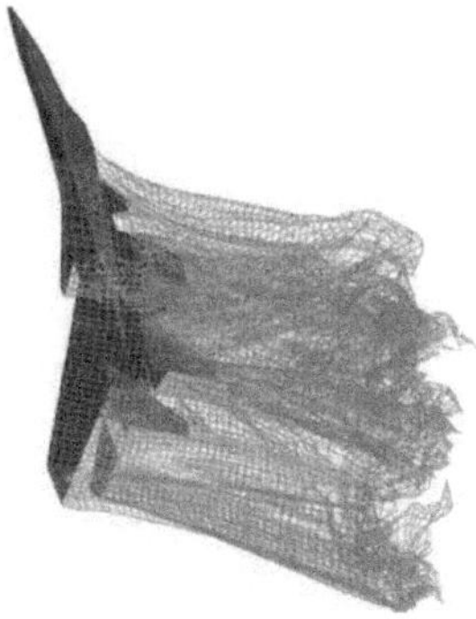

Fig. 3.10. Near vortex wake behind the Su-30 aircraft

The near vortex wake of the A-340 aircraft at $\alpha = 7°$ can be seen in Fig. 3.9. Thin plate representation is used for its lifting and control surfaces, the fuselage and engine nacelles are modeled with solid elements. The wing's high-lift devices are in takeoff configuration.

In computing the near vortex wake behind the Su-30 fighter at $\alpha = 76°$ (Fig. 3.10), the lifting and control surfaces along with the fuselage were modeled with thin plates.

Chapter 4

FAR VORTEX WAKE BEHIND A TURBOJET AIRCRAFT

4.1. The algorithm for computation of the far vortex wake behind aircraft

The algorithm for computing the far vortex wake behind turbojet aircraft is presented as a block diagram in Fig. 4.1. The basis for the algorithm is a mathematical model of the far vortex wake described below. This model is supplemented with initial data from the mathematical model of the near vortex wake in the form of the wake's track left in the control plane (CP) located behind the aircraft perpendicular to its velocity vector at distance 0.5L, where L is the aircraft's length (Fig. 4.2).

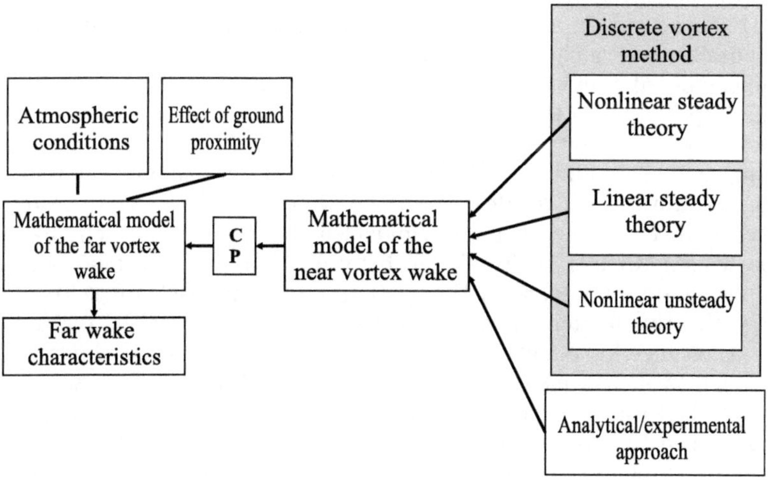

Fig. 4.1. . Block diagram of the algorithm for computation of the far vortex wake (CP - control plane)

All vortices involved in the model are replaced with vortices of infinite length passing through the same points and having the same circulation as with the original vortices. This makes it possible to go on from here to solving a plane problem to which the mathematical model of the far vortex wake is reduced. The mathematical model of the near wake is constructed

A.S. Ginevsky, A.I. Zhelannikov, *Vortex Wakes of Aircrafts*, Foundations of Engineering Mechanics, DOI 10.1007/978-3-642-01760-5_4, © Springer-Verlag Berlin Heidelberg 2009

on the basis of linear steady, nonlinear steady, and nonlinear unsteady theories. These theories are based on the discrete vortex method.

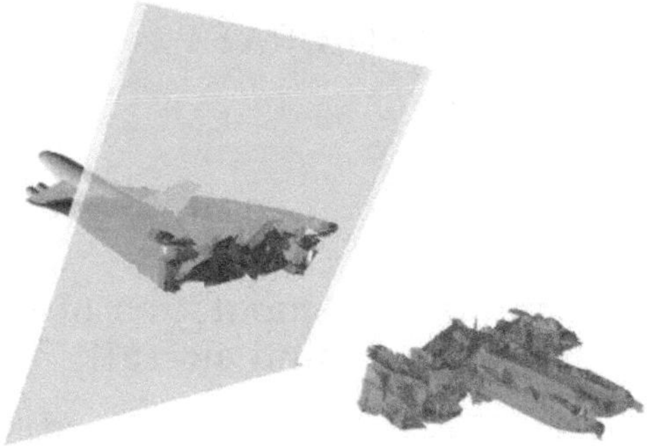

Fig. 4.2. An example of the location of the control plane behind a wake-generating aircraft.

The mathematical model of the near vortex wake may be constructed through the use of an analytical/experimental approach based on Zhukovsky's lift theorem. In this case, the aircraft is replaced by a Π-shaped vortex and having equated the aircraft's lift to its weight it is possible to evaluate the circulation of the wing-tip vortices. Thereupon the computed circulation is redistributed between the wing and horizontal tail proportional to their areas. As a result, four finite tip vortices behind the aircraft are obtained along with the coordinates of the tips of the wing and horizontal tail. If the aircraft is in the landing configuration, the shedding of six vortices from the wing (two from its tips and four from both ends of each flap) is modeled. In this case the circulation of the wing's vortices is distributed between all the six vortices according to the relative flapspan. In Ref. [83], such an approach turned out to be very effective in computing wakes behind aircraft of various types.

Allowance for the effect of jet engines' plumes is provided in the same manner as it was made in works [7, 22, 24, 45]. Engine power setting is specified through the use of an actuator section, whereas the plume is modeled in the course of the development of the vortex sheet shed from the exit shroud. The atmospheric conditions are taken into account with the Richardson number (see below). In the present book, an atmospheric states rating scale is introduced (see Section 4.6). The ground proximity is taken into account in the manner used in works [8, 30]. With the vortices approaching the ground, a near-wall flow is generated on the ground surface, accompanied with boundary-layer separation and formation of secondary vortices. Their interaction with the primary vortices leads to alteration of the latters' motion and to the so-called vortex rebound. These vortices can also be taken into consideration (see the corresponding methodology in Chapter 7). As a result, we obtain the characteristics of the far vortex wake.

4.2. Mathematical model of the far vortex wake

The mathematical model of the far vortex wake is constructed on the assumption that the near wake has been computed with the discrete vortex method and the circulations of bound and free vortices are known. The basis for the far wake mathematical model is the exact solution of the Helmholtz equation, which for planar flow has the following form:

$$\frac{\partial \Omega_x}{\partial t} + W_y \frac{\partial \Omega_x}{\partial y} + W_z \frac{\partial \Omega_x}{\partial z} = \nu \left(\frac{\partial^2 \Omega_x}{\partial y^2} + \frac{\partial^2 \Omega_x}{\partial z^2} \right), \tag{4.1}$$

where Ω_x is a component of the angular velocity vector $\overrightarrow{\Omega}$; W_y and W_z are components of the linear velocity vector of fluid flow $\overrightarrow{W}$; ν is the kinematic viscosity coefficient. Equation (4.1) describes the diffusion process of a vortex in a viscous incompressible fluid.

If at the initial instant of time $t = 0$ a vortex with circulation Γ_{+i} is located at the point (y_i, z_i) and is parallel to the $0x$ axis, according to the exact solution of equation (4.1), for any arbitrary point on the $z0x$ plane we have

$$\Omega_{x(i)}(y, z, t) = \frac{\Gamma_{+i}}{8\pi\nu} \frac{1}{t} e^{-\frac{(y-y_i)^2+(z-z_i)^2}{4\nu t}}. \tag{4.2}$$

In accordance with the Stokes theorem and taking into account the initial condition $\Gamma_{+i}(y, z, 0) = \text{const}$, the expression for the circulation to be determined at the point (x, z) at any arbitrary instant of time t will be written as

$$\Gamma_{+i}(y, z, t) = \Gamma_{+i}(y, z, 0) \left[1 - e^{-\frac{(y-y_i)^2+(z-z_i)^2}{4\nu t}} \right]. \tag{4.3}$$

The components of the velocity vector for this point at any instant of time are determined by the expressions

$$W_{y(i)} = \frac{\Gamma_{+i}}{2\pi} \frac{z - z_i}{(y - y_i)^2 + (z - z_i)^2} \left[1 - e^{-\frac{(y-y_i)^2+(z-z_i)^2}{4\nu t}} \right], \tag{4.4}$$

$$W_{z(i)} = \frac{\Gamma_{+i}}{2\pi} \frac{y - y_i}{(y - y_i)^2 + (z - z_i)^2} \left[1 - e^{-\frac{(y-y_i)^2+(z-z_i)^2}{4\nu t}} \right],$$

or, in a nondimensional form,

$$\overline{W}_{y(i)} = \frac{\Gamma_i}{2\pi} \frac{\overline{z} - \overline{z}_i}{(\overline{y} - \overline{y}_i)^2 + (\overline{z} - \overline{z}_i)^2} \left[1 - e^{-\mathrm{Re}\frac{(\overline{y}-\overline{y}_i)^2+(\overline{z}-\overline{z}_i)^2}{4\tau}} \right], \tag{4.5}$$

$$\overline{W}_{z(i)} = \frac{\Gamma_i}{2\pi} \frac{\overline{y} - \overline{y}_i}{(\overline{y} - \overline{y}_i)^2 + (\overline{z} - \overline{z}_i)^2} \left[1 - e^{-\mathrm{Re}\frac{(\overline{y}-\overline{y}_i)^2+(\overline{z}-\overline{z}_i)^2}{4\tau}} \right].$$

In expressions (4.5), $\Gamma_i = \Gamma_{+i}/U_0 b$, $\overline{y} = y/b$, $\overline{z} = z/b$, $\tau = U_0 t/b$, $\mathrm{Re} = U_0 b/\nu$, U_0 is the undisturbed flow velocity.

Thus, the foundation for the mathematical model is formulas (4.5) containing, instead of the Reynolds number and the coefficient ν, their corrected counterparts Re^* and ν^*. The way of obtaining the coefficient ν^* is described Chapter 7. The coefficient ν^* can be evaluated through processing experimental data. If the location of the vortex wake behind the vortex-generating aircraft is known, the coefficient ν^* can be computed with the following formula derived through processing experimental data and the application of Zhukovsky's lift theorem [8]:

$$\nu^* = -\frac{U_0 l_0^2}{4}\left[\frac{1}{N}\sum_{i=1}^{N} x_i \ln\left(1 - \frac{16 l_0^2 \operatorname{tg}\varphi_i}{C_{ya} S}\right)\right]^{-1}, \tag{4.6}$$

where U_0 is the aircraft's speed, $l_0 = \pi l/4$, l is the wing span, C_{ya} is the lift coefficient defined as $C_{ya} = C_{ya}^{\alpha}\alpha$, ($\alpha$ — angle of attack), S is the wing area, x_i is the distance from the aircraft to the cross section i, where the wake's loss of height is measured; φ_i is the angle between the aircraft's flight path and the direction to the vortex' track in the ith control cross section; N is the number of the control sections where the wake's height is measured.

4.3. Check for the existence and uniqueness of the solution

Consider an airfoil moving in an incompressible fluid medium with kinematic viscosity ν at constant speed U_0 and angle of attach α.

Let the $Oxyz$ frame of reference be attached to the airfoil. In the case under consideration the flow circulation around the airfoil varies, and, according to the theorem on the conservation of circulation about a fluid contour not traversing singularities, free vortices will emanate from the airfoil and move with the flow with their circulation remaining constant in the framework of the given mathematical model. In this case the free vortices will be parallel to the bound vortices on the airfoil, i.e., to the $0z$ axis. The medium's vorticity is taken into consideration through the velocity induced in the medium by a vortex of unit intensity, which makes it possible to meet the condition for circulation to be constant in the space more accurately.

Since the problem is considered in a linear formulation for low angles of attack, free vortices are assumed to move in the Oxy plane at velocity U_0.

Let the airfoil section occupy the segment $[-1, 1]$ of the Ox axis. Because of the fact that the free vortex sheet leaves the trailing edge of the airfoil, the velocities in the sheet must be finite and, consequently, the control point must be located closer to the trailing edge, whereas the discrete vortex must be closer to the leading edge.

The arrangement of discrete vortices and control points must be such as to allow the discrete vortices to be located at points

$$x_i = -1 + \left(i - \frac{3}{4}\right) h, \quad h = \frac{2}{n}, \quad i = \overline{1, n}, \tag{4.7}$$

and the control points to be

$$x_{0i} = x_i + \frac{h}{2} = -1 + \left(i - \frac{1}{4}\right) h, \quad i = \overline{1, n}. \tag{4.8}$$

Let us observe the time evolution of the entire vortex sheet at intervals $\Delta t = h/U_0$. For simplicity let $U_0 = 1$. The coordinate of a free vortex left the airfoil at instant t_s will be $\xi_{sr} = x_n + (r - s + 1)\,\Delta t$ at the current instant t_r. Let the circulation of the discrete vortex at point x_i be Γ_{ir} at the current instant of time t_r, and circulations of free vortices left the airfoil until this point be Λ_s^r, $s = 1, \ldots, r$ and time invariant. With the tangency condition met at the airfoil's control points x_{0i}, $i = 1, \ldots, n$, we have the system of equations

$$\sum_{i=1}^{n} \Gamma_{ir}\omega_{ij} + \sum_{s=1}^{r} \Lambda_s^r \omega_{sjr} = -V_j^*, \quad j = \overline{1, n}, \quad r = 1, 2, \ldots, \tag{4.9}$$

where V_j^* is the normal component of the free-stream flow velocity at control point x_{0j}, ω_{ij} is the normal component of the velocity at control point x_{0j}, induced by the vortex of unit intensity located at point x_i; ω_{sjr} is the normal velocity component at point x_{0j}, induced by the vortex of unit intensity at point ξ_{sr} in the viscous medium, with

$$\omega_{sjr} = \frac{1 - e^{-(t_r - t_s)^2/4\nu t_r}}{x_{0j} - x_n - (t_r - t_s + \Delta t)}, \tag{4.10}$$

following from Ref. [27].

Let us supplement the system under consideration with an equation representing a discrete form of the circulation's constancy in the entire space. If the airfoil is set into motion from rest, this equation has the form

$$\sum_{i=1}^{n} \Gamma_{ir} + \sum_{s=1}^{r} \Lambda_s^r = 0, \quad r = 1, 2, 3, \ldots. \tag{4.11}$$

Thus, we have the following system of equations:

$$\begin{aligned} &\sum_{i=1}^{n} \Gamma_{ir}\omega_{ij} + \sum_{s=1}^{r} \Lambda_s^r \omega_{sjr} = -V_j^*, \qquad j = \overline{1, n}, \quad r = 1, 2, \ldots, \\ &\sum_{i=1}^{n} \Gamma_{ir} + \sum_{s=1}^{r} \Lambda_s^r = 0, \qquad\qquad\qquad\qquad\qquad r = 1, 2, 3, \ldots. \end{aligned} \tag{4.12}$$

Setting $\Gamma_{ir} = \gamma(x_i, t_r)\,h$, $\Lambda_s^r = \delta(t_s)\,\Delta t$, we can rewrite this system as

$$\sum_{i=1}^{n} \frac{\gamma(x_i,t_r)\,h}{x_{0j}-x_i} + \sum_{s=1}^{r} \frac{\delta(t_s)\left(1-e^{-(t_r-t_s)^2/4\nu t_r}\right)}{x_{0j}-x_n-(t_r-t_s+\Delta t)}\Delta t = 2\pi f(t_r),$$
$$j=\overline{1,n}, \quad r=1,2,3\ldots, \tag{4.13}$$
$$\sum_{i=1}^{n} \gamma(x_i,t_r)\,h + \sum_{s=1}^{r} \delta(t_s)\,\Delta t = 0, \quad r=1,2,3,\ldots, \quad f(t_r) = -V_j^*.$$

Since the matrix of this system is non-degenerate, it follows that the system is solvable [27]. On the other hand, according to the results of work [22], let us assume that the functions $\gamma(x,t)$ and $\delta(t)$ belong to the class H^* on the corresponding sets. As a result, we find that the preceding system approximates the following system of integral equations:

$$\int_{-1}^{1} \frac{\gamma(x,t)}{x_0-x}dx + \int_{0}^{t} \frac{\delta(\tau)\left(1-e^{-(t-\tau)^2/4\nu t}\right)}{x_0-1-(t-\tau)}d\tau = 2\pi f(t), \quad x_0\in[-1,1], \quad t\geqslant 0$$
$$\int_{-1}^{1} \gamma(x,t)\,dx + \int_{0}^{t} \delta(\tau)\,d\tau = 0, \tag{4.14}$$

where for the H^* class functions with $\varphi(t) \in H^*$, $t\in[a,b]$

$$\varphi(t) = \frac{\varphi(t)}{(t-a)^{\alpha}(b-t)^{\mu}}, \quad \varphi(t) \in H \text{ на } [a,b], \quad 0 \leqslant \alpha,\ \mu \leqslant 1. \tag{4.15}$$

We shall show that system (4.14) has a unique solution meeting the required boundary conditions. For physical reasons, the first equation of system (4.14) will be solved for the unbounded function $\gamma(x,t)$ at point $x=-1$ as a singular integral equation of index $k=0$ on the interval [−1,1].

By solving this equation for $\gamma(x,t)$, we get

$$\gamma(x,t) = 2\sqrt{\frac{1-x}{1+x}}\,f(t) + \frac{1}{\pi^2}\sqrt{\frac{1-x}{1+x}}\int_{0}^{t}\delta(\tau)\left(1-e^{-(t-\tau)^2/4\nu t}\right)d\tau\,\times$$
$$\times\int_{-1}^{1}\sqrt{\frac{1+x_0}{1-x_0}}\,\frac{dx_0}{(x-x_0)(x_0-1-(t-\tau))}. \tag{4.16}$$

The following formulas from Ref. [27] will be used for further manipulations:

$$\int_{-1}^{1} \sqrt{\left(\frac{1+x}{1-x}\right)} \frac{dx}{x_0 - x} \equiv -\pi, \quad \int_{-1}^{1} \sqrt{\left(\frac{1-x}{1+x}\right)} \frac{dx}{x_0 - x} \equiv \pi, \quad |x_0| \leqslant 1,$$

$$\int_{-1}^{1} \frac{dx}{(x-b)\sqrt{1-x^2}} = \pm \frac{\pi}{\sqrt{b^2-1}}, \tag{4.17}$$

where the minus sign is for $b > 1$ and the plus sign for $b < 1$.

From the latter integral we find

$$\int_{-1}^{1} \sqrt{\frac{1+x_0}{1-x_0}} \frac{dx_0}{b - x_0} = -\pi + \pi\sqrt{\frac{b+1}{b-1}}, \qquad b \geqslant 1,$$

$$\int_{-1}^{1} \sqrt{\frac{1-x}{1+x}} \frac{dx}{b - x} = \pi - \pi\sqrt{\frac{b-1}{b+1}}, \qquad b \geqslant 1. \tag{4.18}$$

Using formulas (4.17) and (4.18), we get

$$\int_{-1}^{1} \sqrt{\frac{1+x_0}{1-x_0}} \frac{dx_0}{(x-x_0)(x_0 - 1 - (t-\tau))} = \sqrt{\frac{2+t-\tau}{t-\tau}} \frac{\pi}{1+(t-\tau)-x}. \tag{4.19}$$

Then, relation (4.16) takes the form

$$\gamma(x,t) = 2\sqrt{\frac{1-x}{1+x}} f(t) + \tag{4.20}$$

$$+\frac{1}{\pi}\sqrt{\frac{1-x}{1+x}} \int_0^t \sqrt{\frac{2+t-\tau}{t-\tau}} \frac{\left(1 - e^{-(t-\tau)^2/4\nu t}\right)}{1+t-\tau-x} \delta(\tau) d\tau.$$

Upon substituting expression (4.20) into the second equation of system (4.8), this equation will take the form

$$\int_0^t \sqrt{\frac{2+t-\tau}{t-\tau}} \delta(\tau)\, d\tau + \int_0^t \left(1 - \sqrt{\frac{2+t-\tau}{t-\tau}}\right) \times$$

$$\times e^{-(t-\tau)^2/4\nu t} \delta(\tau)\, d\tau = 2\pi f(t), \quad t \in [0, T]. \tag{4.21}$$

Let $f \in C_1[0,T]$ and $f(0)=0$. Then every solution of equation (4.20) continuous on the interval $0 < t \leqslant T$ satisfies the integral equation:

$$\delta(t) + \int_0^t k(t,\tau)\,\delta(\tau)\,d\tau = 2\pi f'(t), \quad 0 < t \leqslant T, \tag{4.22}$$

$$k(t,\tau) = \left(e^{-(t-\tau)^2/4\nu t} - 1\right)\frac{1}{\sqrt{2+t-\tau}\,(t-\tau)^{3/4}} +$$
$$+ e^{-(t-\tau)^2/4\nu t}\,\frac{\sqrt{t-\tau}\,(t+\tau)}{2\nu t^2\left(\sqrt{2+t-\tau}+\sqrt{t-\tau}\right)},$$

which is obtained from (4.21) by termwise differentiating equation with respect to t. It is easy to see that the opposite is also true: every solution of equation (4.22) continuous on $0 < t \leqslant T$ satisfies equation (4.21).

By multiplying both sides of equation (4.22) by $\sqrt{t}$ and introducing the new unknown function $y(t) = \delta(t)\sqrt{t}$, we get

$$y(t) + 2\int_0^{\sqrt{t}} \sqrt{t}\,k\left(t,x^2\right) y\left(x^2\right) dx = 2\pi f'(t)\sqrt{t}. \tag{4.23}$$

By virtue of the fact that there are two inequalities,

$$\left|2\sqrt{t}\,k\left(t,x^2\right)\right| \leqslant M, \quad 0 \leqslant x^2 \leqslant t \leqslant T \quad \text{and} \quad \left|2\pi f'(t)\sqrt{t}\right| \leqslant N, \tag{4.24}$$

a solution of equation (4.23) exists and is unique. It can be written in the form of the sum of the following series:

$$y(t) = y_0(t) - y_1(t) + y_2(t) - \ldots, \quad y_0(t) = 2\pi f'(t)\sqrt{t},$$
$$y_{k+1}(t) = \int_0^{\sqrt{t}} 2\sqrt{t}\,k\left(t,\tau^2\right) y_k\left(\tau^2\right) d\tau, \quad k = 0,1,2,\ldots. \tag{4.25}$$

Since equation (4.23) is a corollary of expression (4.21), a solution of the original equation exists and has the form:

$$\delta(t) = \frac{1}{\sqrt{t}}\sum_{i=0}^{\infty}(-1)^i y^i(t). \tag{4.26}$$

From this follows the existence and uniqueness of the solution of the systems of equations (4.20) and (4.21), i.e., the following theorem holds true.

Theorem. The solution of the system of equations

$$\gamma(x,t) = 2\sqrt{\frac{1-x}{1+x}}\, f(t) + \frac{1}{\pi}\sqrt{\frac{1-x}{1+x}}\int_0^t \sqrt{\frac{2+t-\tau}{t-\tau}}\,\frac{\left(1-e^{-\frac{(t-\tau)^2}{4\nu t}}\right)}{1+t-\tau-x}\,\delta(\tau)\,d\tau,$$

$$\int_0^t \sqrt{\frac{2+t-\tau}{t-\tau}}\,\delta(\tau)\,d\tau + \int_0^t \left(1-\sqrt{\frac{2+t-\tau}{t-\tau}}\right) e^{-\frac{(t-\tau)^2}{4\nu t}}\,\delta(\tau)\,d\tau = 2\pi f(t),$$

$$t \in [0,T], \quad f(0)=0, \quad f(t) \in C_1[0,T], \tag{4.27}$$

exists and is unique.

This theorem provides the justification for the above model of flow around a thin airfoil.

4.4. Similarity considerations for flow in the far vortex wake

A great diversity of aircraft types makes the problem of studying far vortex wakes difficult and multivariant. In the course of investigations into the characteristics of the far trailing wake behind aircraft, it was found that the wakes have many common features. It turned out that if certain conditions are met, it is possible to provide flow similarity in different far trailing wakes. Let us identify the corresponding similarity criteria.

Consider the process of the formation of the vortex wake behind a thin rectangular wing of moderate aspect ratio. Assume that the flow past the wing's leading edge is smooth and attached. The vortex sheet leaving the side and trailing edges rolls up into two vortex tubes and, spreading behind the wing, sinks. Of special interest is the distance from the trailing edge where the sheet can be considered fully rolled up. The process of the wake's transformation takes place along the entire space behind the wing, which is why the notion «fully rolled-up sheet» has an asymptotic meaning and for determining the distance for the sheet to become rolled up it is necessary to use additional particular criteria based, for example, on the assessment of computational accuracy or the specified level of errors for computed flow parameters. The assessment of the indicated distance can be performed using approaches of slender-body theory. From similarity theory, the distance L for the sheet to become fully rolled up can be defined with the formula

$$L = k\frac{bV_0}{V'}, \tag{4.28}$$

where k is a correction factor, b is the characteristic length (wing chord), V_0 is the velocity of undisturbed flow, V' is the velocity induced by the wing's vortex system.

The velocity V' is determined by the value of the velocity circulation Γ_0 around the wing section, and, as a consequence,

$$L \approx \frac{bb_L V_0}{\Gamma_0}, \tag{4.29}$$

where b_L is the length of the vortex with circulation Γ_0, which is tied to the wing's lift Y_a with the relation подъемной силой крыла Y_a соотношением

$$Y_a = \rho_0 V_0 \Gamma_0 b_L. \tag{4.30}$$

From (4.29) and (4.30) it follows that

$$L \approx \frac{bb_L^2 \rho_0 V_0}{Y_a} \approx b\frac{\lambda}{C_{ya}}, \tag{4.31}$$

where λ is the wing's aspect ratio. Since the characteristic linear dimension b for the wake behind the wing is proportional to its span l, we get from (4.31)

$$\frac{L}{l} = k\frac{\lambda}{C_{ya}}. \tag{4.32}$$

The coefficient k in (4.32) is determined with regard to the roll-up of the vortex sheet.

It is evident that in the sheet's rolling up into two vortex tubes the circulation Γ_0 of each tube is equal to the circulation for the wing's root section and, in accordance with the Zhukovsky theorem, is related to the lift coefficient C'_{ya} for this section and its chord b by the formula

$$\Gamma_0 = \frac{V_0}{2}(C'_{ya} b). \tag{4.33}$$

To determine the distance l_0 between the two vortex tubes let us consider the system of bound vortices modeling the wing as a combination of П-shaped vortices with one branch of each П-vortex lying in the root section. For such a representation of the wing, its lift is determined by the relation

$$Y_a = \rho_0 V_0 \sum_{i=1}^{n} \Gamma_i z_i, \tag{4.34}$$

where z_i is distance from the root chord to the vortices modeling the vortex sheet, n is the number of the sheet's vortices, each with circulation Γ_i.

From expression (4.34) it follows that for the lift constant and no slip the z coordinate of the vortex combination centroid, z_c, for each wing is also constant and determined with the formula

$$z_c = \frac{1}{\sum_{i=1}^{n_0} \Gamma_i} \sum_{i=1}^{n_0} \Gamma_i z_i, \tag{4.35}$$

where n_0 is the number of vortices modeling the half-wing.

With no ground effect, $z_c = \text{const}$ in each wake cross section and thus is equal to the lateral coordinate of the vortex tube:

$$z_c = \frac{l_0}{2}. \tag{4.36}$$

Taking into account that

$$\sum_{i=1}^{n_0} \Gamma_i = \Gamma_0 \sum_{i=1}^{n_0} \Gamma_i z_i = \frac{Y_a}{2\rho V_0}, \tag{4.37}$$

expression (4.36) changes to

$$l_0 = \frac{Y_a}{\rho_0 V_0 \Gamma_0}. \tag{4.38}$$

Using expression (4.33), we find

$$l_0 = \frac{Y_a}{\frac{\rho_0 V_0^2}{2} C'_{ya} b} = \frac{C_{ya}}{C'_{ya}} \frac{S}{b}. \tag{4.39}$$

Relation (4.39) is derived in the framework of assumptions of wing linear theory, in particular this is related to the definition of the wing's lift. From slender body theory, a simple formula for l_0 can be derived:

$$l_0 = \frac{\pi}{4} l. \tag{4.40}$$

It should be noted that relation (4.40) can be obtained in the framework of linear theory through the use of an additional assumption about the elliptic circulation distribution along the wing span l. Using expressions (4.34), (4.35) and (4.39), it is possible to find criteria of wings' equivalence from the standpoint of similarity of disturbances introduced into the flow by the wings:

$$(C'_{ya} b)_1 = (C'_{ya} (C_{ya} b)_1 = (C_{ya} b)_2, \tag{4.41}$$

where subscripts 1 and 2 denote quantities related to the two wings being compared.

It is evident that the above relationships are also true for a system of lifting surfaces symmetric relative to a plane passing through the velocity vector of undisturbed flow. Being applied to a system of m lifting surfaces, formulas (4.23) and (4.39) take the form

$$\Gamma_0 = \frac{V_0}{2} \sum_{i=1}^{m} (C'_{ya} b)_i, \quad l_0 = \frac{\sum_{i=1}^{m} (C_{ya} S)_i}{\sum_{i=1}^{m} (C'_{ya} b)_i}, \tag{4.42}$$

where i is the number of a lifting surface.

It follows from relations (4.33), (4.39) and (4.42) that vortex wakes with coinciding basic parameters can be generated by different wings and lifting systems. Consequently, l_0 and Γ_0 can be considered as similarity parameters for aerodynamic surfaces in terms of vortex wake characteristics. For flat-plate monoplane wings with simple planforms characterized by the aspect ratio λ, leading edge sweep χ and taper ratio η, the equivalence conditions with respect to wake parameters, i.e., equivalence relations, may be presented in the form

$$\begin{aligned} \frac{l_k}{l} &= \frac{C_{ya}^{\alpha} C_{yk}^{'\alpha}}{C_{yak}^{\alpha} C_{y}^{'\alpha}} \left[\frac{1+\eta^{-1}}{1+\eta_k^{-1}}\right]^2; \\ \frac{b_k}{b} &= \frac{C_{ya}^{\alpha} C_{yk}^{'\alpha}\lambda}{C_{yak}^{\alpha} C_{y}^{'\alpha}\lambda_k} \left[\frac{1+\eta^{-1}}{1+\eta_k^{-1}}\right]^2; \\ \frac{\alpha_k}{\alpha} &= \frac{C_{yak}^{\alpha}\lambda_k}{C_{ya}^{\alpha}\lambda} \left[\frac{C_{y}^{'\alpha}}{C_{yk}^{'\alpha}}\right] \left[\frac{1+\eta^{-1}}{1+\eta_k^{-1}}\right]^2. \end{aligned} \qquad (4.43)$$

The quantities with no subscript in (4.43) relate to a certain basic wing, those with subscribes $k = 1, 2, \ldots$ are quantities related to a family of equivalent wings.

Thus, when studying the characteristics of aircraft vortex wakes, it is possible by using similarity criteria (4.41) to reduce the problem in hand to consideration of wake characteristics and disturbances generated by an equivalent wing. Such an approach provides a significant simplification of studying, for example, the effect of the far vortex wake on the flight dynamics of aircraft, since the use of similarity criteria with respect to the vortex wake characteristics decreases the number of variants to be considered.

As an example, Fig. 4.3 demonstrates fields of disturbance velocities behind the Tu-154M aircraft (*a*) and an equivalent wing (*b*) at a downstream distance of $X = 153\,\mathrm{m}$. One can see a significant discrepancy between these fields.

The disturbance velocity fields at a distance of $X = 4132\,\mathrm{m}$ are shown in Fig. 4.4. In this case the vortex wakes behind the Ty-154M (*a*) and its equivalent wing (*b*) are fully identical.

4.5. A universal procedure for transition to the mathematical model of the far vortex wake

To perform the transition to modeling the far vortex wake of different aircraft, using the disturbance velocity fields just behind them as initial data, a special universal procedure was developed. The disturbance velocity field in the control plane behind the aircraft under consideration is obtained by mathematically modeling the flow past the aircraft through nonlinear unsteady theory. The coordinates of points and components of disturbance velocities at the points in the control plane serve as the initial data for the transition procedure. In this case the velocity components become boundary conditions for determining unknown circulations of new

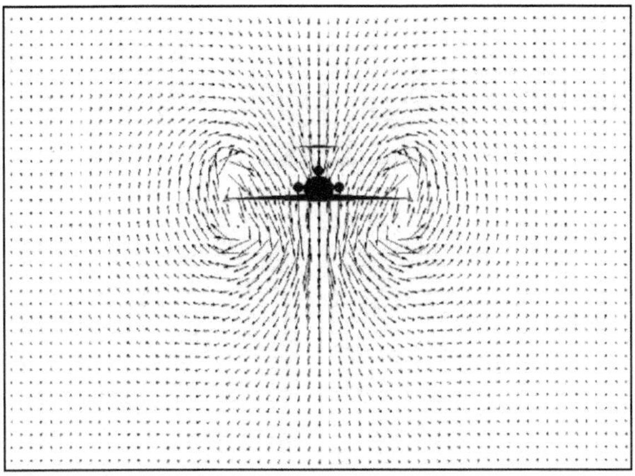

a

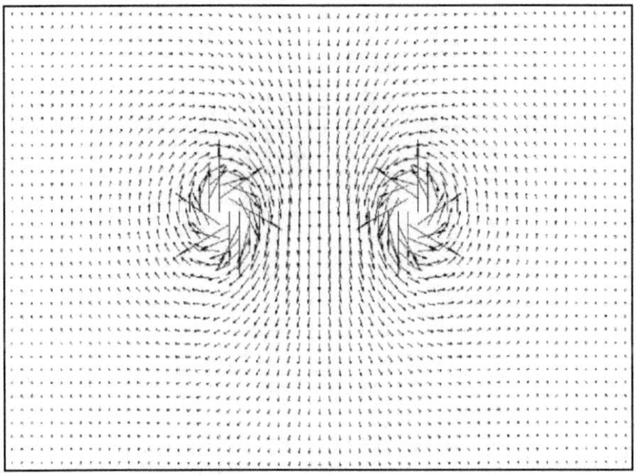

b

Fig. 4.3. Disturbance velocity fields behind the Tu-154M (*a*) aircraft and the equivalent wing (*b*) at a downstream distance of $X = 153\,\mathrm{m}$

rectilinear infinite vortices located in the same plane. Intensities of the indicated vortices must be such that the disturbance velocities induced by them at control points be equal to the velocities induced at the same points by the aircraft. The problem reduces to solving a system of linear algebraic equations of the type

$$\sum_{\mu=1}^{2n} \Gamma_{\mu} a_{z,y\nu\mu} = 2\pi w_{z,y\nu}; \quad \nu = 1, n, \tag{4.44}$$

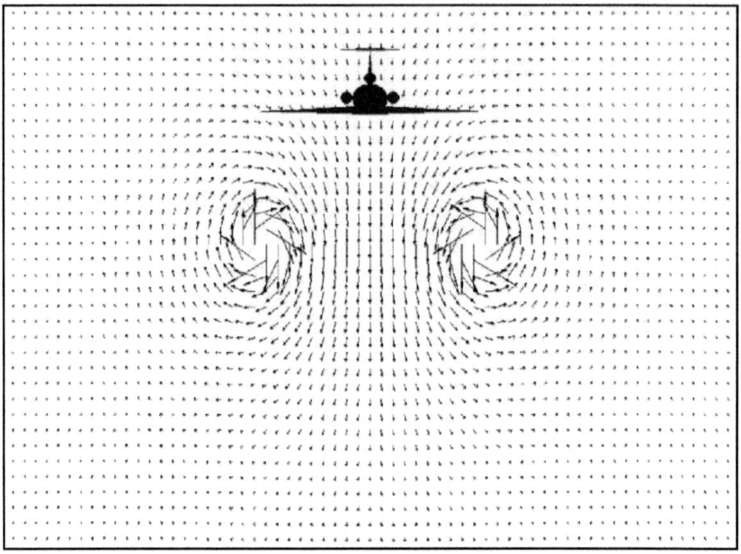

a

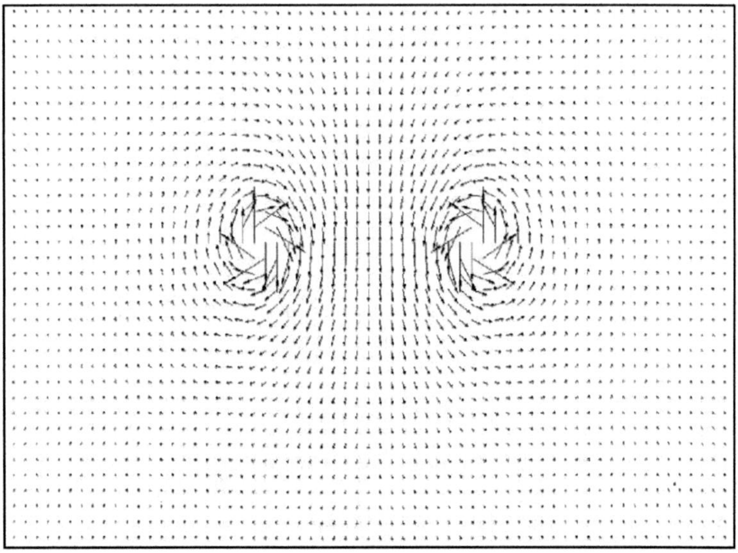

b

Fig. 4.4. Disturbance velocity fields behind the Tu-154*M* aircraft (*a*) and the equivalent wing (*b*) at a downstream distance of $X = 4132\,\mathrm{m}$

where the components of the disturbance velocity $w_{(z,y)}$ and nondimensional coefficients $a_{(z,y)}$ are known quantities, whereas the intensities of the rectilinear infinite vortices Γ_μ are sought-for quantities.

The number of the vortices, μ, in (4.44) must be twice the number of the points ν (Fig.4.5), at which velocity fields were determined, which is required for regularization of the system of linear algebraic equations. Having found in such a way the intensities of the rectilinear infinite vortices and having the coordinates of their centers in the specified plane, it is possible to go to the mathematical model of the far vortex wake. To check for operability of the universal procedure, vortex structures behind the An-26 aircraft were computed using nonlinear unsteady theory. Used as hydrodynamic singularities in this case were vortex frames.

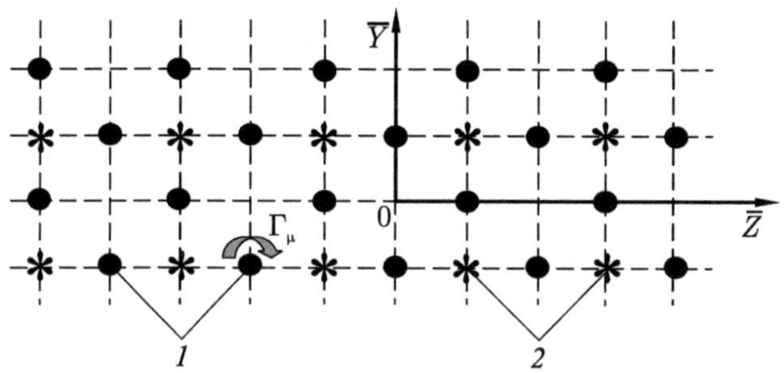

Fig. 4.5. Computational grid with control points and vortices for transition to a mathematical model of the far vortex wake

Fig. 4.6 demonstrates velocity fields behind the An-26 aircraft in a control cross section (at a downstream distance equal to the aircraft's length). Presented at the top the figure is the original velocity field obtained by nonlinear unsteady theory, the lower portion shows the velocity field after replacing vortex frames with rectilinear vortices of infinite length. It is seen that the fields are fully identical. Therefore, such a replacement is rightful.

4.6. Consideration of the state of the atmosphere

For assessment of atmospheric conditions it is convenient to use the Richardson number

$$\mathrm{Ri} = \frac{g}{\overline{T}} \frac{\Delta\overline{T}/\Delta z}{(\Delta N/\Delta z)^2}. \tag{4.45}$$

In formula (4.45), g is the free-fall gravity acceleration; $\overline{T}$ is the mean temperature of the layer of Δz thick; $\Delta\overline{T}/\Delta z$ and $\Delta\overline{N}/\Delta z$ are the corresponding mean gradients of temperature and horizontal wind velocity within the layer Δz in thickness.

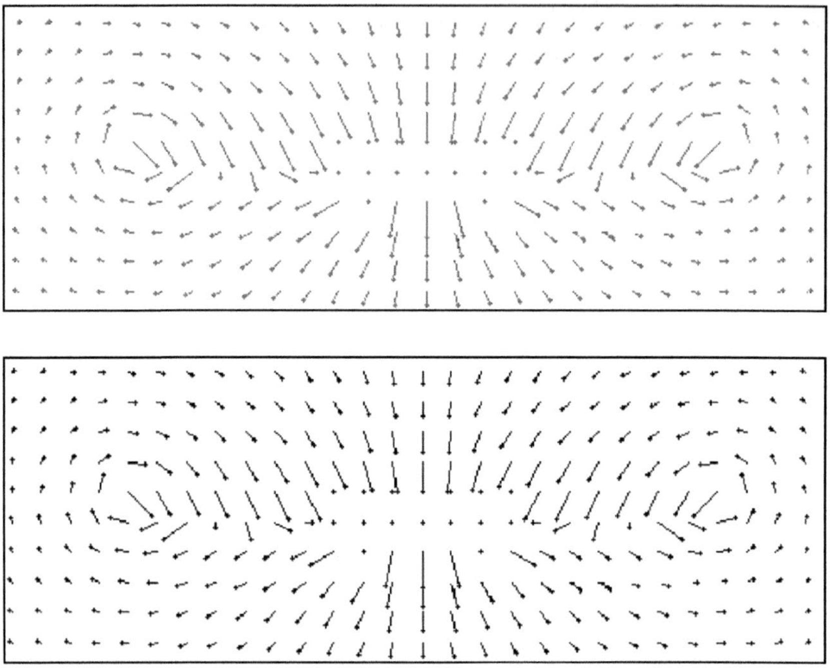

Fig. 4.6. Velocity field behind the An-26 aircraft

The Richardson number characterizes the relation between the buoyancy force (numerator) and the dynamic factor (denominator), i.e., the ratio of contributions of free and forced convections in formation of atmospheric turbulence. So, an increase in the magnitude of the temperature gradient corresponds to the state where buoyancy forces prevail. Increasing the velocity gradient corresponds to an increase in the dynamic factor and characterizes the atmosphere as unstable. An atmospheric state is considered neutral at $-0{,}01 \leqslant \mathrm{Ri} \geqslant 0{,}01$; in this case the thermal effect is minimal and only forced convection can exist. For lower Ri numbers, $\mathrm{Ri} < -0{,}01$, buoyancy forces begin to play a greater role and mixed convection arises; at $\mathrm{Ri} < -1{,}0$ the state of free convection takes place. And conversely, at Richardson numbers exceeding 0,01, buoyancy forces begin to hinder the development of turbulence. At $\mathrm{Ri} > 0{,}25$, the flow becomes near-laminar, with turbulent mixing nearly absent. Thus, the above reasoning can be summarized in a scale for rating atmospheric conditions with condition numbers (CNs) by assigning a numerical value from 1 to 5 to each atmospheric states considered here. A strongly unstable atmosphere is evaluated with number 5, a stable atmosphere with number 1. All other intermediate atmospheric conditions fall within the range.

Richardson number Ri	Atmospheric conditions	Condition number (CN)
$\mathrm{Ri} < -1{,}0$	Very unstable	5
$-0{,}01 > \mathrm{Ri} \geqslant -1{,}0$	Unstable	4
$0{,}01 \geqslant \mathrm{Ri} \geqslant -0{,}01$	Neutral	3
$2{,}5 \geqslant \mathrm{Ri} > 0{,}01$	Stable	2
$\mathrm{Ri} > 0{,}25$	Very stable	1

Flight experiments show that the lifespan of vortex wakes behind aircraft is strongly dependent on atmospheric conditions. In a calm atmosphere vortex wakes live long, whereas in a rough atmosphere they rapidly decay. In the mathematical model of the far vortex wake a relation is used between the atmospheric rating numbers and the corrected Reynolds number, obtained by processing experimental data, which allows one to assess the lifespan of vortex wakes.

The above table agrees with Ref. [72] containing a similar gradation of atmospheric conditions.

4.7. Verification of the method and predicted results

To verify the methodology's credibility and the reliability of data obtainable with it, calculations were performed to compare their results with flight experiments. This section contains examples of such calculations for jet aircraft. Fig. 4.7 demonstrates computational results for the vortex wake behind the An-124 (diamonds) and the data of a flight experiment (squares) obtained by the O. K. Antonov design bureau and presented by V. Komov. One can see a satisfactory agreement between theory and experiment.

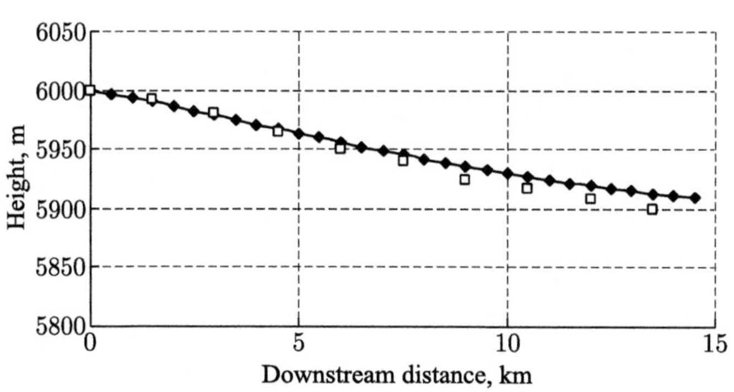

Fig. 4.7. Descent of the vortex wake behind the An-124 aircraft

In Fig. 4.8 is shown the computed descent of the vortex tubes behind the Il-76 aircraft. These data taken from Ref. [8] represent the conditions

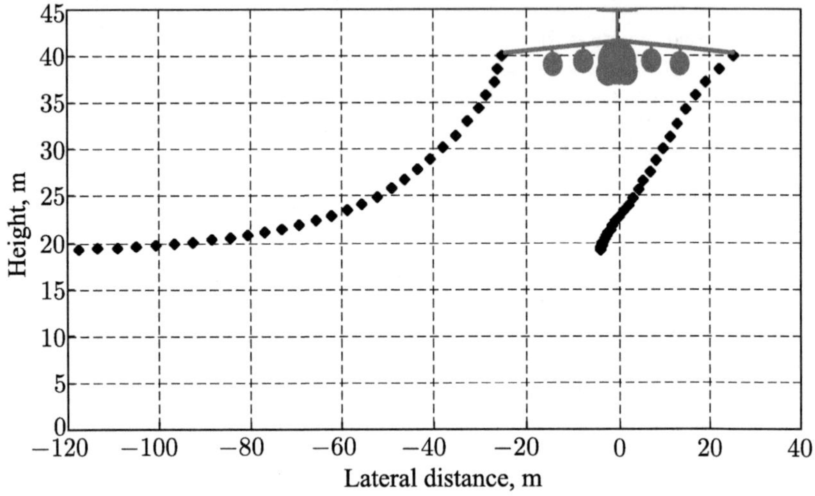

Fig. 4.8. Tracks of the wing-tip vortices in the control plane behind an Il-76 aircraft with the effect of a crossw

of the crash of a Jak-40 aircraft on January 16, 1987 in Tashkent (in the former USSR). An investigation had shown that the cause of the crash was an Il-76's right wing-tip vortex, which, under the action of a crosswind of 0,5–1,5 m/s from the right, had drifted to the centerline of the runway and hanged over it at a height of 20 m.

The Il-76 had passed the site of the subsequent crash at a height of 40 m. Computations based on the proposed methodology show that under these conditions the right tip vortex 20 m must really hang over the centerline of the runway, which also confirms the operability of the methodology used.

Fig. 4.9. presents the vertical velocity distribution in the vortex core of a Il-76 60 sec after its passage over the point where the aircraft's speed at a height of 100 m was measured (squares) during an experiment at the Gromov Flight Research Institute. The computed data are also shown on the same graph (diamonds). The agreement between theory and experiment may be considered satisfactory.

Shown in Fig. 4.10 is the vertical velocity distribution (squares) in the core of the vortex behind a B-747 aircraft 10 sec after its passage at a height of 100 m over the measurement station (experimental data from the Department of Transportation, USA). Data calculated with the proposed methodology are also plotted on the graph (diamonds). Here again a satisfactory agreement is observed between theory and experiment.

Thus, the operability of the methodology developed is confirmed by the reliable results obtained with it.

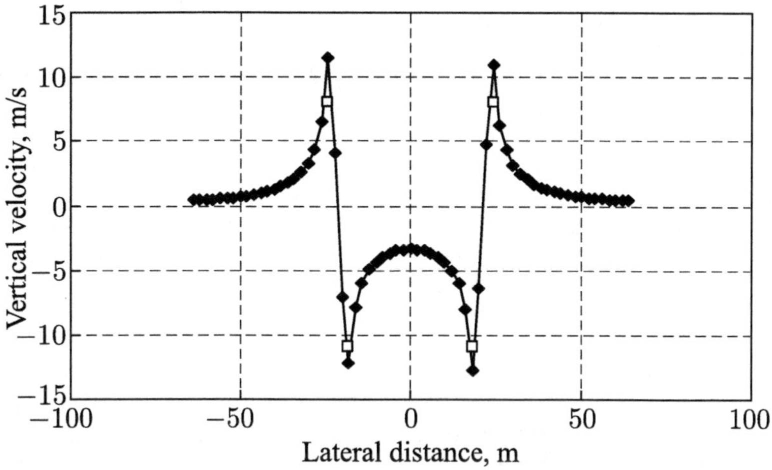

Fig. 4.9. Vertical velocity distribution in the core of the wing-tip vortex of the Il-76 aircraft

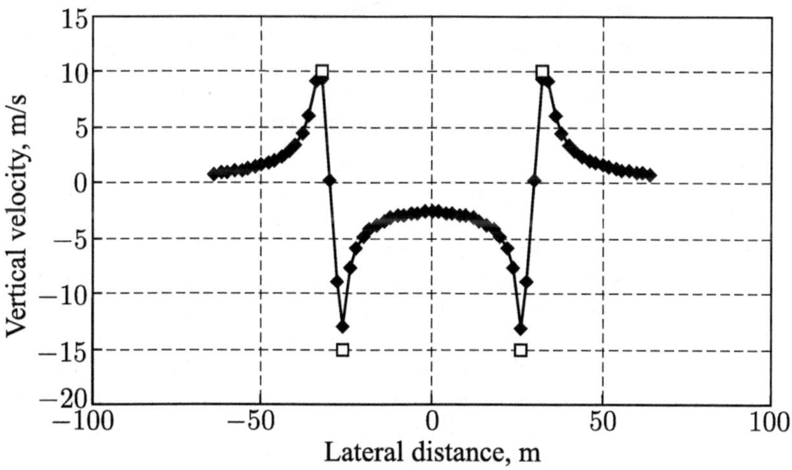

Fig. 4.10. Vertical velocity distribution in the vortex core of a B-747

4.8. The characteristics of the vortex wake behind the Il-76 aircraft

Using the methodology considered above, wake characteristics of different aircraft were investigated under various flight conditions.

This section deals with the results of studying the characteristics of the far vortex wake behind the Il-76 aircraft; the data obtained have allowed one to reveal the effect of the flight speed and altitude on the position of the vortex wake as well as the influence of atmospheric conditions considered in section 4.6.

The computed data presented in Fig. 4.11 demonstrate the descent of the vortex tubes behind the Il-76 flying at a height of $H = 1{,}000$ m and a speed of $V = 300$ km/h for different atmospheric conditions: very stable (CN= 1), neutral (CN= 3) and very unstable (CN= 5). It can be seen that in a very stable atmosphere the vortex wake at a distance of $X = 15$ km from the aircraft sank $\Delta H = -110$ m. In a very unstable atmosphere the vortex wake rapidly attenuates due to vortex diffusion and the loss of height of the vortex tubes is only 30 m.

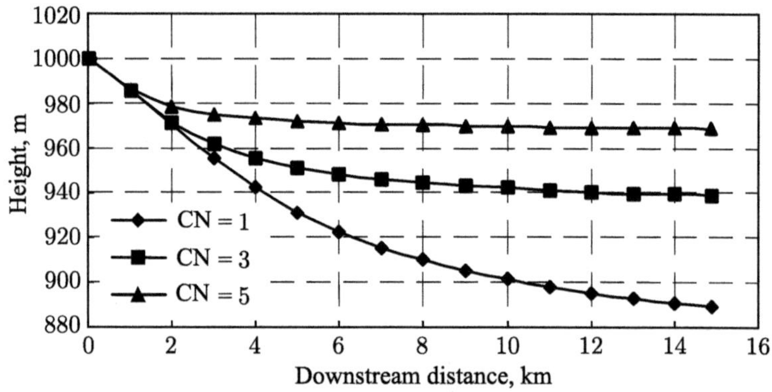

Fig. 4.11. Descent of the vortex tubes behind the Il-76 aircraft flying at ($V = 300$ km/h, $H = 1{,}000$ m) m under different atmospheric conditions

Similar results in Fig. 4.12 are presented for the Il-76 aircraft flying at $H = 1{,}000$ m and $V = 500$ km/h under the same atmospheric conditions. It is seen that in a very stable atmosphere (CN= 1) the vortex wake sank $\Delta H = -55$ m at a distance of $X = 15$ km behind the aircraft. In a very unstable atmosphere (CN= 5) the vortex wake sank $\Delta H = -15$ m.

Fig. 4.13 demonstrates the computational results for the Il-76 aircraft flying at a height of $H = 1{,}000$ m and speed of $V = 700$ km/h under the same atmospheric conditions. As may be seen from the graph, in a very calm atmosphere (CN= 1) the vortex wake sank $\Delta H = -35$ m at a distance of $X = 15$ km behind the aircraft. Under very unstable atmospheric conditions (CN= 5) the vortex wake lost $\Delta H = -12$ m in height.

Thus, it was found that with increasing speed at the same flight height the descent of the vortex tubes decreases due to decreasing their circulation.

With increasing the flight height, the situation begins to change. Fig. 4.14 demonstrates computational results for the vortex wake behind the Il-76 aircraft flying at a height of $H = 5{,}000$ m and speed of $V = 300$ km/h at different atmospheric conditions: CN= 1, 3, and 5.

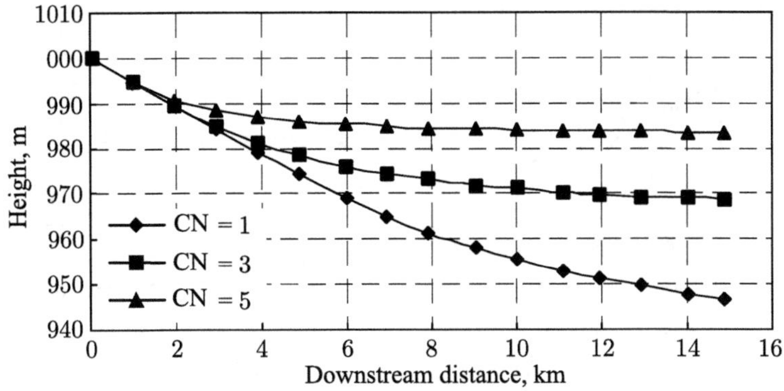

Fig. 4.12. Descent of the vortex tubes behind the Il-76 aircraft flying at ($V = 500\,\text{km/h}$, $H = 1{,}000\,\text{m}$) under different atmospheric conditions

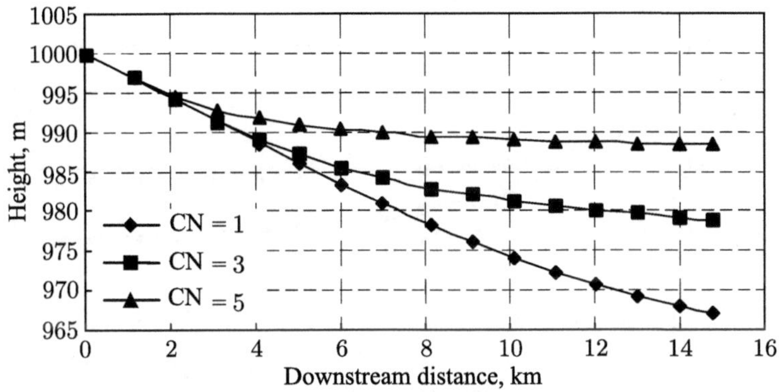

Fig. 4.13. Descent of the vortex tubes behind the Il-76 aircraft flying at ($V = 700\,\text{km/h}$, $H = 1{,}000\,\text{m}$) under different atmospheric conditions

It is seen that in a very calm atmosphere (CN= 1) the loss of height of the vortex wake at a distance of $X = 15\,\text{km}$ behind the aircraft is $\Delta H = -150\,\text{m}$. In a very unstable atmosphere (CN= 5), the vortex wake sank $\Delta H = -40\,\text{m}$. Compare: for the aircraft flying at $H = 1{,}000\,\text{m}$ the descent is $\Delta H = -110\,\text{m}$ at CN= 1 and $\Delta H = -30\,\text{m}$ at CN= 5 (see Fig. 4.11), i.e., with increasing flight height, all other factors equal, the wake's sinking increases. A similar situation can be observed for other flight speeds (Figs. 4.13, 4.14).

Presented in fig. 4.15 are the computational results for the descent of the vortex tubes of the Il-76 aircraft flying at $H = 5{,}000\,\text{m}$ and $V = 500\,\text{km/h}$ under different atmospheric conditions. In a very stable atmosphere the vortex wake at $X = 15\,\text{km}$ behind the aircraft sank 70 m. In a very unstable atmosphere the wake's descent was 25 m. Compare: at

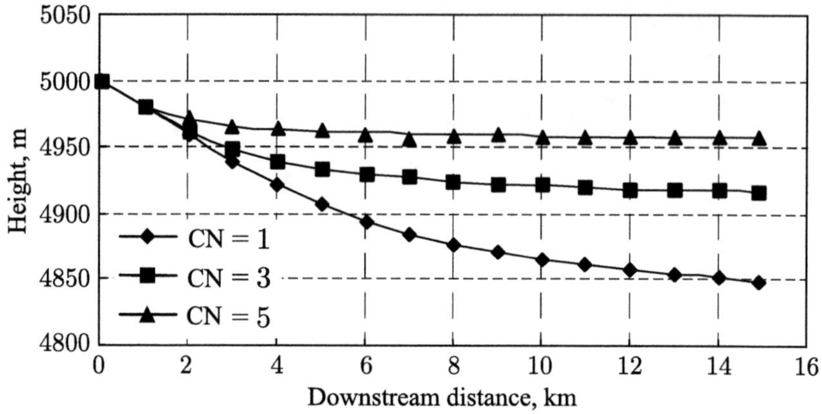

Fig. 4.14. Descent of the vortex wake behind the Il-76 aircraft flying at ($V = 300\,\text{km/h}$, $H = 5{,}000\,\text{m}$) under different atmospheric conditions

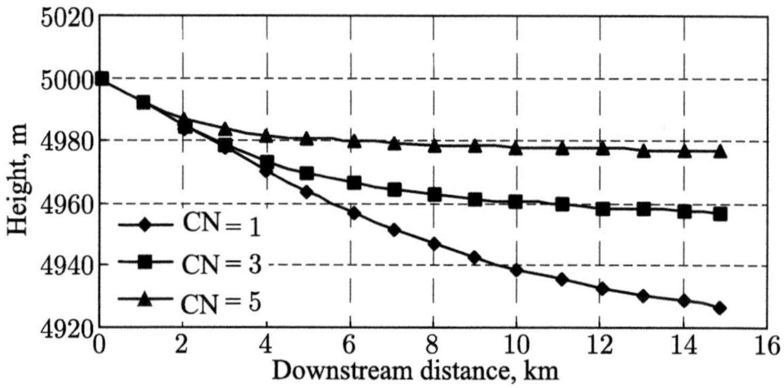

Fig. 4.15. Descent of the vortex wake behind the Il-76 aircraft flying at ($V = 500\,\text{km/h}$ and $H = 5{,}000\,\text{m}$) under different atmospheric conditions

$H = 1{,}000\,\text{m}$ and CN= 1 $\Delta H = -55\,\text{m}$, whereas at CN= 5 $\Delta H = -15\,\text{m}$ (see Fig. 4.12).

Fig. 4.16 presents the computed descent of the vortex tubes of the Il-76 aircraft flying at $H = 5{,}000\,\text{m}$ and $V = 700\,\text{km/h}$ under the same atmospheric conditions.

At CN= 1 the descent of the vortex wake at a distance of $X = 15\,\text{km}$ behind the aircraft was $\Delta H = -45\,\text{m}$. At CN= 5 the wake's loss of height at the same point was $\Delta H = -15\,\text{m}$. Compare: 15 km downstream of the aircraft flying at $H = 1{,}000\,\text{m}$ $\Delta H = -35\,\text{m}$ at CN= 1 and $\Delta H = -12\,\text{m}$ at CN= 5 (see Fig. 4.13).

Thus, it was found that with increasing flight height at the same flight speed the descent of the vortex tubes increases due to an increase in their circulation (Figs. 4.14–4.16).

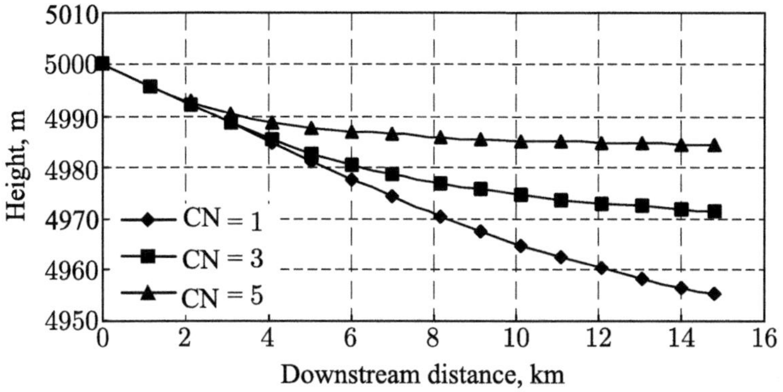

Fig. 4.16. Descent of the vortex plaits behind the Il-76 aircraft flying at (V = 700 km/h H = 5,000 m) under different atmospheric conditions

The effect of atmosphere conditions on the vortices' attenuation can be seen in Figs. 4.17–4.21.

Figs. 4.17 and 4.18 demonstrate computational results for the disturbance velocity fields behind the Il-76 aircraft flying at H = 5,000 m and V = 500 km/h under very unstable atmospheric conditions (CN= 5) at different distances downstream of the aircraft: X = 0; 1,53; 3,06; and 4,58 km. One can see that under such conditions, already at distances 4–5 km behind the aircraft the vortex tubes lose their intensity and pose no threat to other aircraft.

For a calm atmosphere the picture is different. Shown in Figs. 4.19–4.21 are the results of computation of the disturbance velocity fields generated by the Il-76 flying at H = 5,000 m and V = 500 km/h in a very stable atmosphere (CN= 1) at distances of X = 0; 1,53; 3,06; 4,58; 12,5 and 14,86 km behind the aircraft. Under these conditions, the vortex tubes also attenuate, but this takes place at significantly greater distances (about 12–14 km).

4.9. The characteristics of the vortex wake behind the An-124, B-747 and A-380 aircraft

This section deals with the results of studies into the characteristics of the far vortex wake behind the following aircraft: An-124 (wing span L = 73,6 m, takeoff mass G = 405 t), B-747 (L = 59,64 m, G = 365 t) and A-380 (L = 79,8 m, G = 560 t). These aircraft are the world's largest flying machines except for the An-125 «Mriya» aircraft. A comparison of these aircraft's vortex wake characteristics is presented below.

Fig. 4.22 demonstrates the computed descents of the vortex tubes behind the An-124, B-747 and A-380 aircraft flying at H = 1,000 m and V = 300 km/h in a very stable atmosphere (CN= 1). These data show that under very stable atmospheric conditions the vortex wakes at a distance

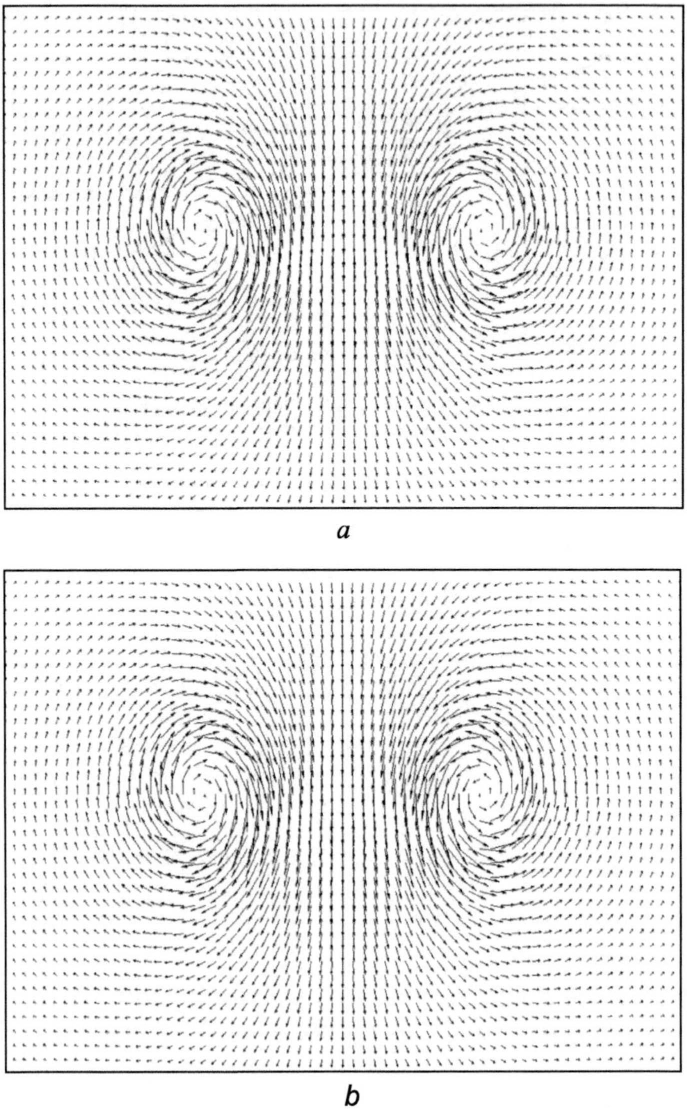

Fig. 4.17. Disturbance velocity fields behind the Il-76 aircraft at $t = 0$ c *(a)* and $t =$ $= 11$ c *(b)* ($V = 500$ km/h $H = 5{,}000$ m, CN$= 5$)

of $X = 15$ km behind the B-747 and A-380 sank 200 m. The descent of the wake behind the An-124 in the same case was 160 m.

Vertical velocities in the core of the wing-tip vortices were also computed. Figs. 4.23–4.25 present the vertical velocity distributions at

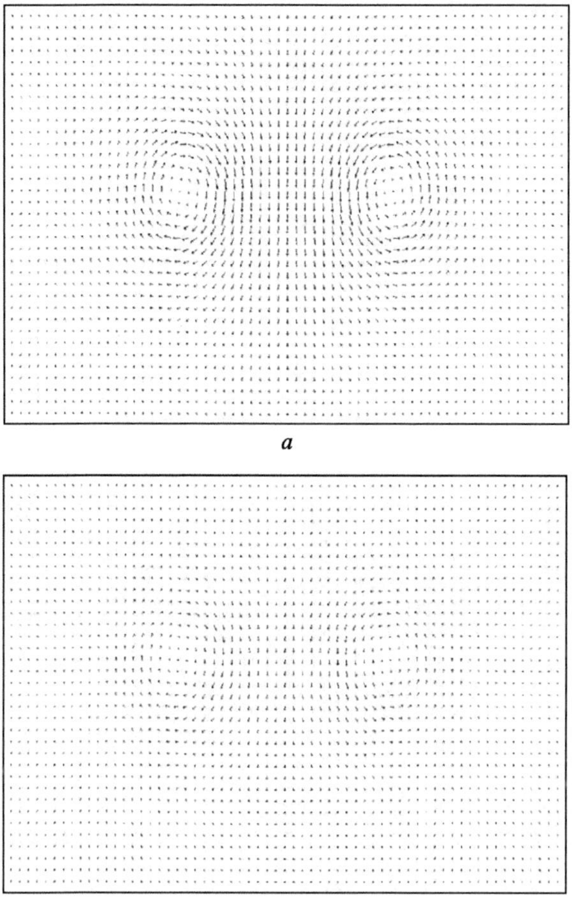

Fig. 4.18. Disturbance velocity fields behind the Il-76 aircraft at $t = 22\,\mathrm{s}$ *(a)* and $t = 33\,\mathrm{s}$ *(b)* ($V = 500\,\mathrm{km/h}$, $H = 5{,}000\,\mathrm{m}$, CN$= 5$)

distances $X = 50\,\mathrm{m}$, $200\,\mathrm{m}$ and $1000\,\mathrm{m}$ behind the An-124, B-747 and A-380 aircraft flying at $H = 1{,}000\,\mathrm{m}$ and $V = 300\,\mathrm{km/h}$, CN$= 1$. In these cases the vertical velocities in the vortex cores of the An-124, B-747 and A-380 at distances of up to $2{,}000\,\mathrm{m}$ do not exceed values within the range of minus $30\,\mathrm{m/s}$ to plus $25\,\mathrm{m/s}$. With increasing distances from the aircraft the vertical velocities decrease and at $X = 10{,}000\,\mathrm{m}$ do not exceed $\pm 12\,\mathrm{m/s}$.

With increasing flight speed the situation becomes different. Fig. 4.26 presents the computed descent of the vortex tubes behind the An-124, B-747 and A-380 aircraft flying at $H = 1{,}000\,\mathrm{m}$ and $V = 500\,\mathrm{km/h}$ at

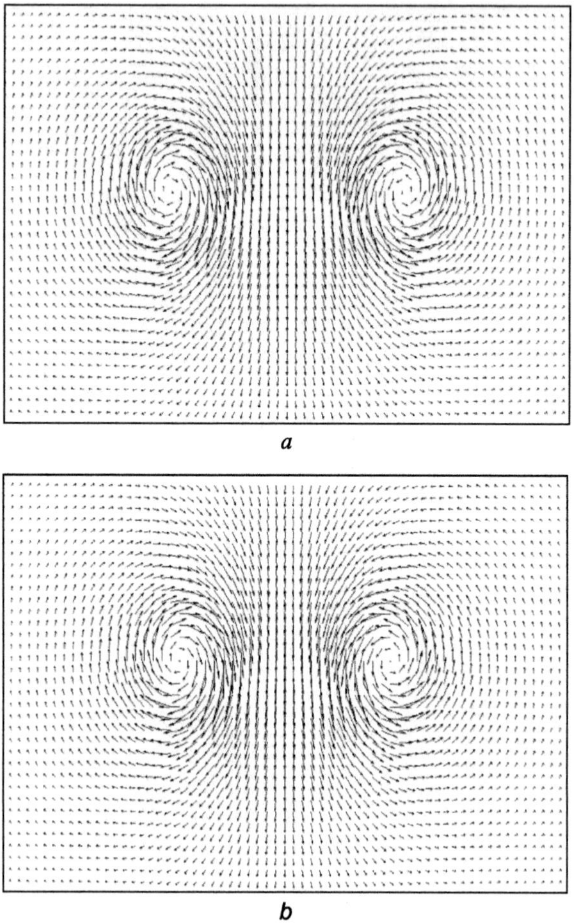

Fig. 4.19. Disturbance velocity fields behind the Il-76 aircraft at $t = 0$ s *(a)* and $t =$ $= 11$ s *(b)* ($V = 500$ km/h, $H = 5{,}000$ m, CN= 1)

CN= 1. The computations have shown that in a very calm atmosphere the vortex wake behind the B-747 and A-380 aircraft at a distance of $X = 15$ km sank 90 m (at $V = 300$ km/h, $\Delta H = -200$ m). At the same distance behind the An-124 $\Delta H = -70$ m (at $V = 300$ km/h, $\Delta H = -160$ m). Thus, the aircraft speed noticeably affects the descent of the vortex wake.

The plots in Figs. 4.27–4.29 show the vertical velocity distributions in the crossplanes 50 m, 2,000 m and 10,000 m downstream of the An-124, B-747 and A-390 aircraft flying at $H = 1{,}000$ m and $V = 500$ km/h, CN= 1. In this case the vertical velocities in the vortex cores of these three aircraft at distances up to 2,000 m do not exceed ± 20 m/s. With

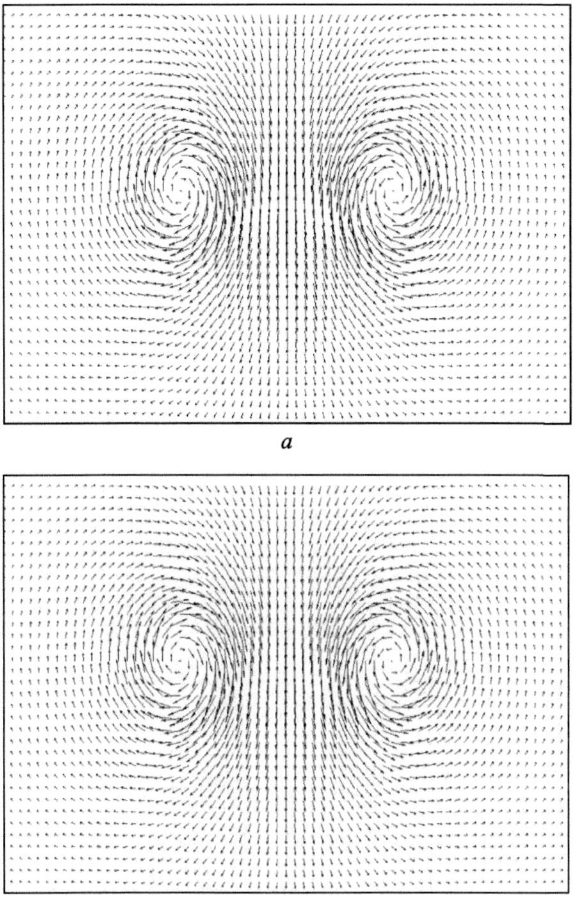

Fig. 4.20. Disturbance velocity fields behind the Il-76 aircraft at $t = 22\,\mathrm{s}$ *(a)* and $t = 33\,\mathrm{s}$ *(b)* ($V = 500\,\mathrm{km/h}$, $H = 5{,}000\,\mathrm{m}$, CN= 1)

increasing the downstream distance, the vertical velocities decrease and at $X = 10{,}000\,\mathrm{m}$ do not exceed $\pm 10\,\mathrm{m/s}$.

Figs. 4.30–4.32 present plots of the vertical velocity distribution at distances $X = 50\,\mathrm{m}$, 2,000 m and 10,000 m behind the An-124, B-747 and A-380, respectively, flying at $H = 1{,}000\,\mathrm{m}$, $V = 500\,\mathrm{km/h}$, CN= 1.

These plots allow one to trace the decay of the vortices with increasing X.

The plots in Fig. 4.33 demonstrate the vertical velocity distribution at a distance of $X = 50\,\mathrm{m}$ behind the An-124, B-747 and A-380 flying at $H = 50\,\mathrm{m}$ and $V = 250\,\mathrm{km/h}$ for CN= 1. Their flaps are in landing position. This situation is typical for landing approach. It can be seen that

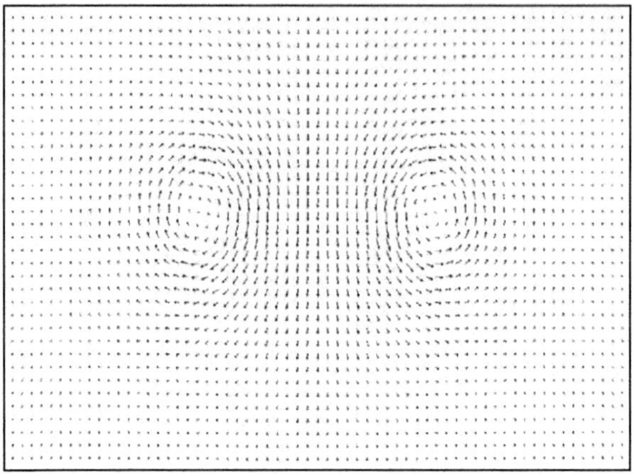

a

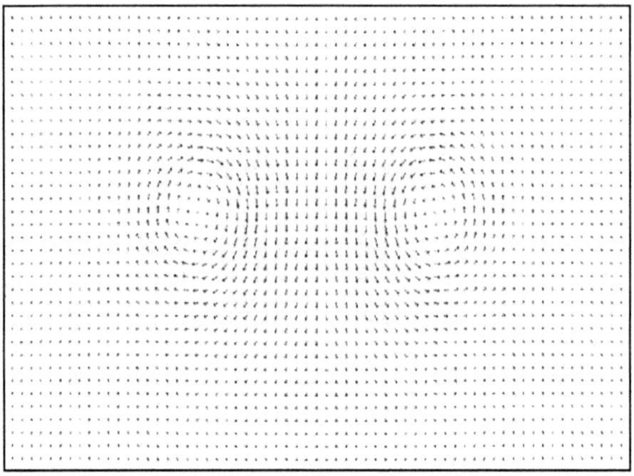

b

Fig. 4.21. Disturbance velocity fields behind the Il-76 aircraft at $t = 90\,\mathrm{s}$ *(a)* and $t = 107\,\mathrm{s}$ *(b)* ($V = 500\,\mathrm{km/h}$, $H = 5{,}000\,\mathrm{m}$, CN= 1)

the B-747 and A-380 aircraft generate vertical velocities of up to $\pm 30\,\mathrm{m/s}$, those for the An-124 are up to $\pm 25\,\mathrm{m/s}$.

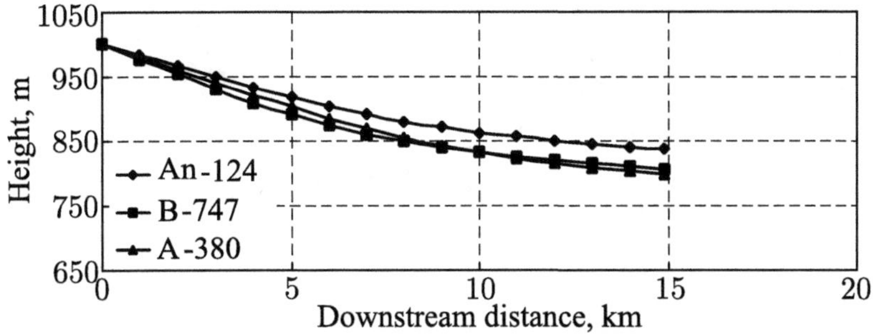

Fig. 4.22. Descent of the vortex plaits behind the An-124, B-747 and A-380 aircraft ($V = 300$ km/h, $H = 1{,}000$ m, CN= 1)

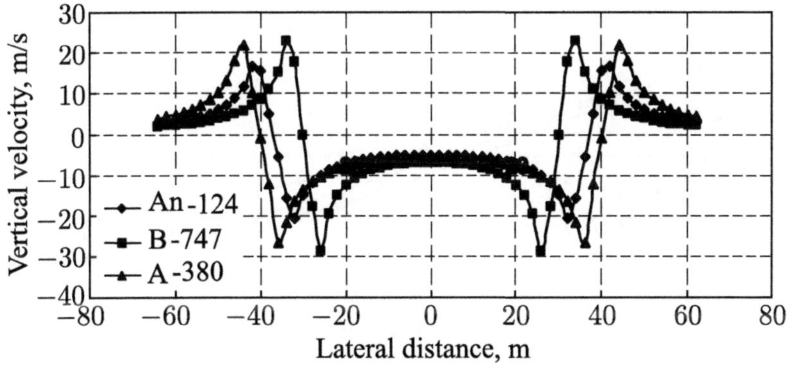

Fig. 4.23. Vertical velocity distribution in the vortex core for the An-124, B-747 and A-380 aircraft ($V = 300$ km/h, $H = 1{,}000$ m, CN= 1, $X = 50$ m)

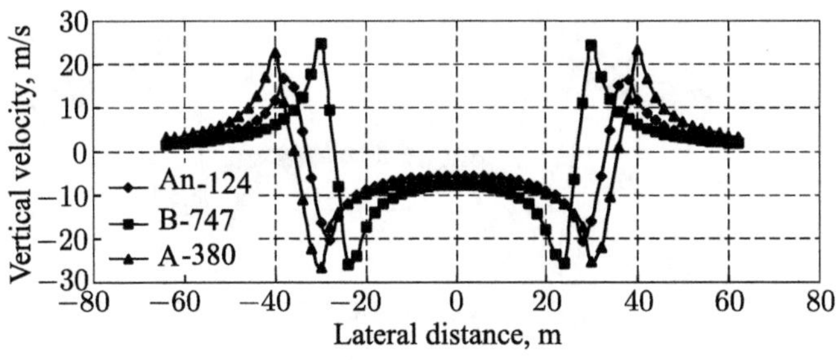

Fig. 4.24. Vertical velocity distribution in the vortex core for the An-124, B-747 and A-380 aircraft ($V = 300$ km/h, $H = 1{,}000$ m, CN= 1, $X = 2{,}000$ m)

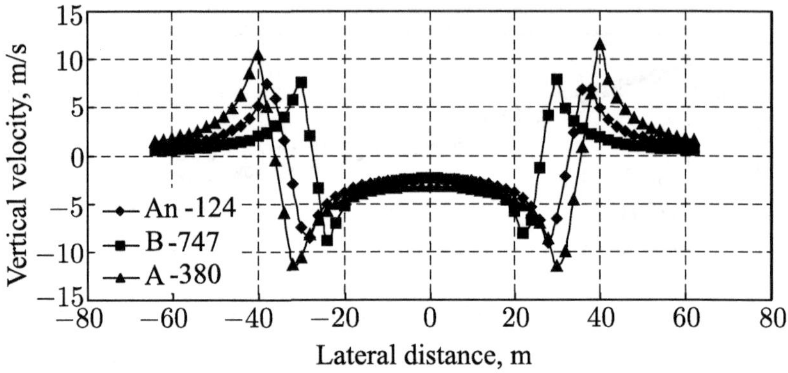

Fig. 4.25. Vertical velocity distribution in the vortex core for the An-124, B-747 and A-380 aircraft B-747 and A-380 ($V = 300$ km/h, $H = 1{,}000$ m, CN= 1, $X = 10{,}000$ m)

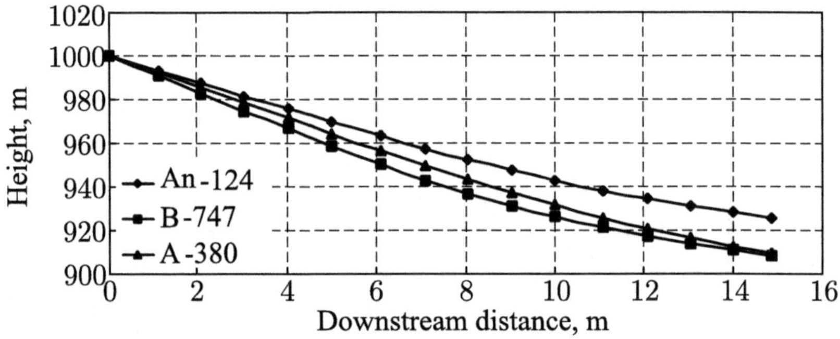

Fig. 4.26. Descent of the vortex wakes behind the An-124, B-747 and A-380 aircraft ($V = 500$ km/h, $H = 1{,}000$ m, CN= 1)

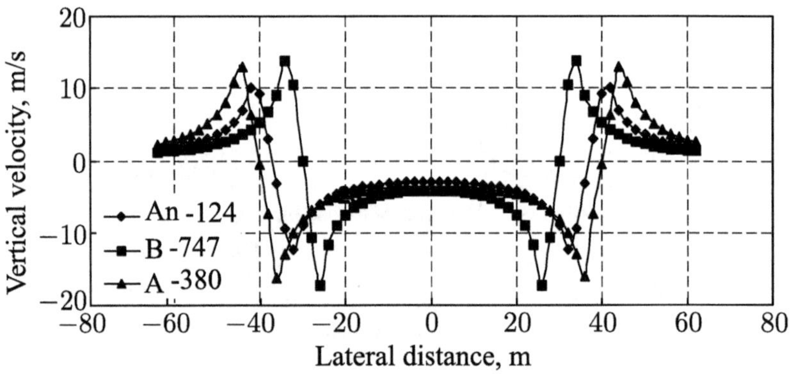

Fig. 4.27. Vertical velocity distributions in the vortex core for the An-124, B-747 and A-380 aircraft ($V = 500$ km/h, $H = 1{,}000$ m, CN= 1, $X = 50$ m)

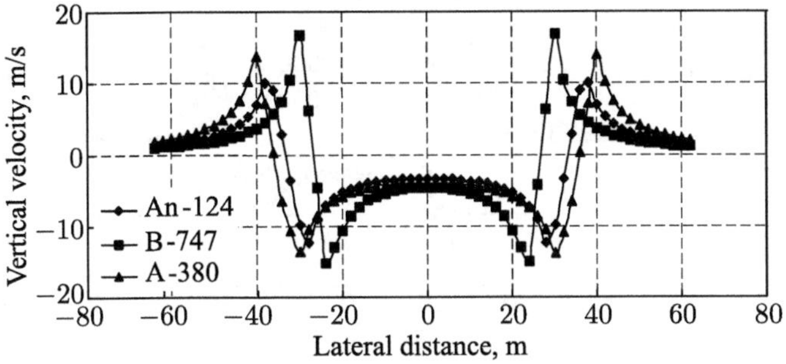

Fig. 4.28. Vertical velocity distributions in the vortex core for the An-124, B-747 and A-380 aircraft ($V = 500\,\text{km/h}$, $H = 1{,}000\,\text{m}$, CN= 1, $X = 2{,}000\,\text{m}$)

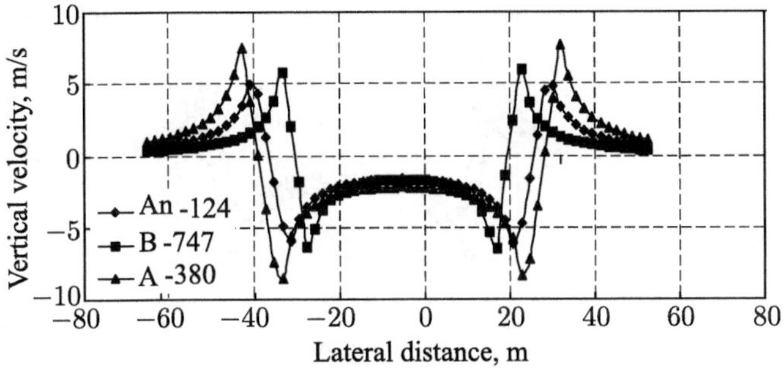

Fig. 4.29. Vertical velocity distributions in the vortex core for the An-124, B-747 and A-380 aircraft ($V = 500\,\text{km/h}$, $H = 1{,}000\,\text{m}$, CN= 1, $X = 10{,}000\,\text{m}$)

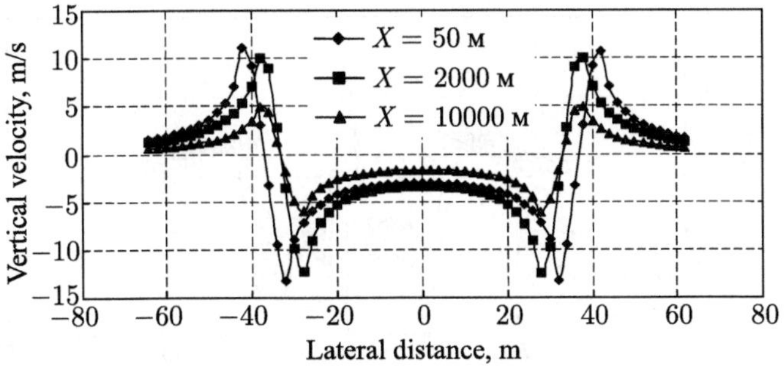

Fig. 4.30. Vertical velocity distributions in the vortex core for the An-124 aircraft ($V = 500\,\text{km/h}$, $H = 1{,}000\,\text{m}$, CN= 1) at different distances behind the generating aircraft

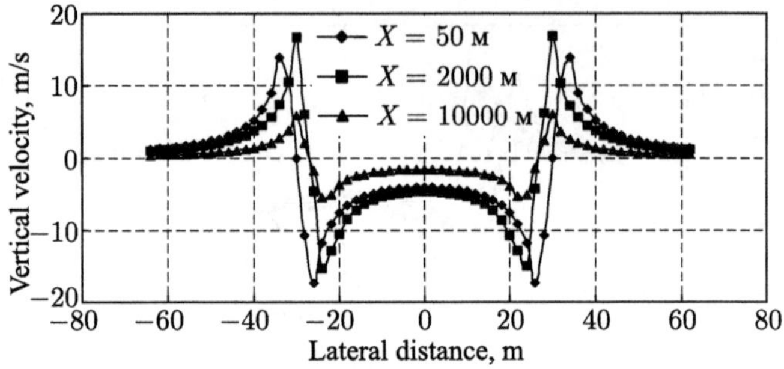

Fig. 4.31. Vertical velocity distributions in the vortex core for the B-747 aircraft ($V = 500$ км/ч, $H = 1{,}000$ m, CN= 1)) at different distances behind the generating aircraft

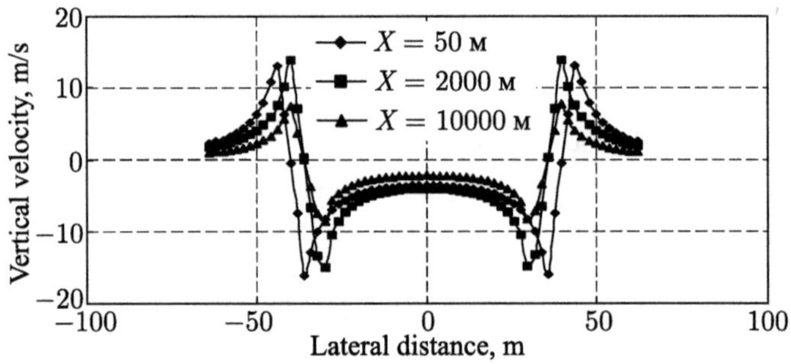

Fig. 4.32. Vertical velocity distributions in the vortex core for the A-380 aircraft ($V = 500$ км/ч, $H = 1{,}000$ m, CN= 1) at different distances behind the generating aircraft

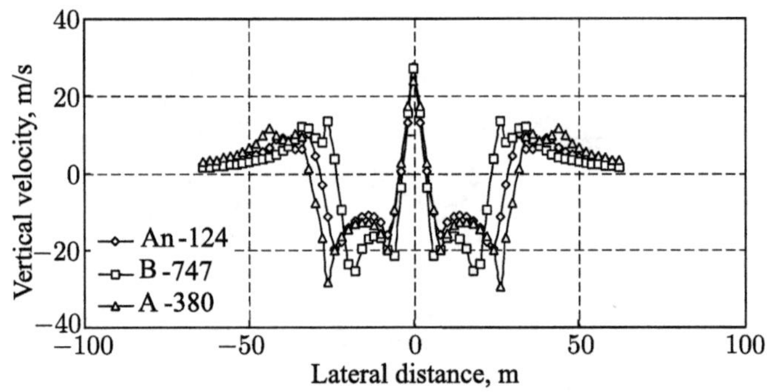

Fig. 4.33. Vertical velocity distributions in the vortex core for the An-124, B-747 and A-380 aircraft ($V = 250$ km/h, $H = 50$ m, CN= 1, $X = 50$ m),) at landing

Chapter 5

VORTEX WAKES BEHIND PROPELLER-DRIVEN AIRCRAFT

5.1. Problem statement

Now we consider the motion of an aircraft with propellers. This aircraft has en route configuration and is traveling with subsonic speed $\vec{W}$ at arbitrary height H.

Let us introduce the following coordinate systems: body axes $Oxyz$, wind axes $OX_a Y_a Z_a$ and earth axes $OX_g Y_g Z_g$ (Fig. 5.1).

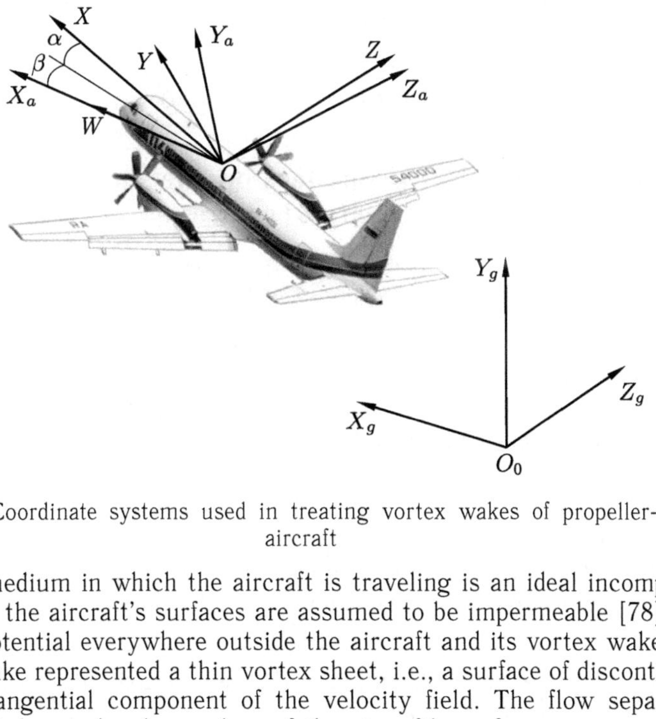

Fig. 5.1. Coordinate systems used in treating vortex wakes of propeller-driven aircraft

The medium in which the aircraft is traveling is an ideal incompressible fluid, the aircraft's surfaces are assumed to be impermeable [78]. The flow is potential everywhere outside the aircraft and its vortex wake. The vortex wake represented a thin vortex sheet, i.e., a surface of discontinuity for the tangential component of the velocity field. The flow separation lines are taken to be sharp edges of the aircraft's surfaces.

The problem reduces to seeking the potential velocity fields $\vec{W}(M,t) = \operatorname{grad} U(M,t)$ and pressure field $p(M,t)$, defined everywhere outside the wing surface σ_1 and the surface σ_2 representing the vortex

A.S. Ginevsky, A.I. Zhelannikov, *Vortex Wakes of Aircrafts*, Foundations of Engineering Mechanics, DOI 10.1007/978-3-642-01760-5_5, © Springer-Verlag Berlin Heidelberg 2009

wake. The vortex wake surface σ_2 is movable, its motion is not known in advance. Here, $\mathrm{M}(x, y, z)$ is a point in space, t is time. In computing potential flows of an ideal incompressible fluid, the incompressibility condition is tantamount to the requirement for the field's potentiality

$$\Delta U\,(M, t) = 0, \tag{5.1}$$

at every point of the space outside the surfaces σ_1 and σ_2, and to satisfy the momentum equation it is sufficient to require meeting the Cauchy-Lagrange condition

$$\frac{P}{\rho} = \frac{P_\infty}{\rho} - \frac{\overline{W}_\infty^2}{2} - \frac{\partial U}{\partial t}, \tag{5.2}$$

where P_∞ is the fluid pressure at infinity, ρ is the fluid density.

On the aircraft's surfaces, the tangency condition is satisfied:

$$\frac{\partial U^\pm}{\partial \overline{n}} = 0, \quad \mathrm{M} \in \sigma_1, \tag{5.3}$$

where $\overline{n}$ is a unit vector perpendicular to the surface σ_1 at the point M. On the vortex sheet at every instant of time, the condition of zero pressure differential is met:

$$p^+(M, t) = p^-(M, t). \tag{5.4}$$

At infinity, the disturbance velocities attenuate:

$$\left|\nabla U\,(M, t) - \overrightarrow{W}_\infty\right| \to 0. \tag{5.5}$$

5.2. The effect of propellers on the far vortex wake characteristics

Let us introduce the following designations:

d — propeller diameter; ω — propeller rotational speed; V_0 — free-stream flow speed;

$V_1 = V_0 + v_1$ — axial stream velocity at the propeller disc (v_1 - corresponding induced velocity);

$V_2 = V_0 + v_2$ — axial stream velocity just behind the propeller (v_2 — corresponding induced velocity);

$v_2 = 2v_1$; — propeller hub radius;

$\xi = \frac{r_0}{R}$ — normalized propeller hub radius,$R = \frac{d}{2}$;

P — propeller thrust; M — propeller shaft torque.

Then $dP = dmV_2 - dmV_1 = dmv_2$, where dP is the thrust element at dr, $dm = 2\pi r dr \rho V_1$ is the mass flow rate at r. Thus,

$$dP = 2\pi\rho V_1 v_2 r dr.$$

By assuming that the distribution of the induced axial velocities over the propeller disc is uniform, we get

$$P = 2\pi\rho V_1 v_2 \int\limits_{r_0}^{R} r dr = 2\pi\rho V_1 v_2 \frac{1}{2}\left(R^2 - r_0^2\right) = \pi\rho V_1 v_2 R^2\left(1 - \xi^2\right).$$

Taking into consideration the equalities $V_1 = V_0 + v_1$ and v_2, we have

$$P = \pi R^2 \rho \left(1 - \xi^2\right) \left(V_0 + v_1\right) 2v_1 = 2\pi R^2 \rho \left(1 - \xi^2\right) \left(V_0 v_1 + v_1^2\right).$$

By introducing the normalized induced axial velocity $\overline{V} = \dfrac{v_1}{\omega R}$, we get

$$P = 2\pi R^2 \rho \left(1 - \xi^2\right) \left(V_0 \omega R \overline{V} + \omega^2 R^2 \overline{V}^2\right).$$

Since $P = \overline{\alpha} \rho n^2 d^4$, $\omega = 2\pi n$ ($\overline{\alpha}$ — propeller thrust coefficient, n — rotational speed),

$$\overline{\alpha} \rho n^2 d^4 = 2\pi \frac{d^4}{4} \rho \left(1 - \xi^2\right) \left(V_0 2\pi n \frac{d}{2} \overline{V} + 4\pi^2 n^2 \frac{d^2}{4} \overline{V}^2\right),$$

or, after factoring ρd^2 out,

$$\overline{\alpha} n^2 d^2 = \frac{\pi}{2} \left(1 - \xi^2\right) \left(V_0 \pi n d \overline{V} + \pi^2 n^2 d^2 \overline{V}^2\right).$$

It follows that

$$\overline{\alpha} = \frac{\pi}{2} \left(1 - \xi^2\right) \left(\frac{V_0 \pi n d}{n^2 d^2} \overline{V} + \frac{\pi^2 n^2 d^2}{n^2 d^2} \overline{V}^2\right).$$

After cancellation, keeping in mind that $\dfrac{V_0}{nd} = \lambda$ (λ — propeller rotational speed factor), we get

$$\overline{\alpha} = \frac{\pi}{2} \left(1 - \xi^2\right) \left(\pi \lambda \overline{V} + \pi^2 \overline{V}^2\right).$$

From this expression we find $\overline{V}$:

$$\pi^2 \overline{V}^2 + \pi \lambda \overline{V} - \frac{2\overline{\alpha}}{\pi \left(1 - \xi^2\right)} = 0,$$

$$\overline{V}^2 + \frac{\lambda}{\pi} \overline{V} - \frac{2\overline{\alpha}}{\pi^3 \left(1 - \xi^2\right)} = 0,$$

$$D = \frac{\lambda^2}{\pi^2} + \frac{8\overline{\alpha}}{\pi^3 \left(1 - \xi^2\right)},$$

$$\overline{V} = \frac{1}{2} \left[-\frac{\lambda}{\pi} + \sqrt{\frac{\lambda^2}{\pi^2} + \frac{8\overline{\alpha}}{\pi^3 \left(1 - \xi^2\right)}}\right] \Big/ 2, \quad \text{or}$$

$$\overline{V} = -\frac{\lambda}{2\pi} + \sqrt{\frac{\lambda^2}{4\pi^2} + \frac{2\overline{\alpha}}{\pi^3 \left(1 - \xi^2\right)}}.$$

Out of the two roots of the quadratic equation, we take the positive one.

For the propeller torque we have the formula $dM = dm u_2 r$, where $u_2(r)$ is the distribution of the circumferential induced velocity over a cross section of the propeller slipstream far behind the propeller disc.

Taking into account the expression $dm = 2\pi r dr \rho V_1$, we get $dM = 2\pi\rho V_1 u_2 r^2 dr$.

Hence it follows that

$$M = 2\pi\rho V_1 \int_{r_0}^{R} u_2(r)\, r^2 dr.$$

The power consumed by the propeller is $N = M\omega$, i.e.,

$$N = M\omega = 2\pi\rho\omega V_1 \int_{r_0}^{R} u_2(r)\, r^2 dr.$$

Due to the adopted vortical representation of the propeller, we can take $u_2(r) = \frac{\Gamma}{2\pi r}$, where Γ — circulation of the axial vortex generated by the propeller. Then

$$\int_{r_0}^{R} u_2(r)\, r^2 dr = \int_{r_0}^{R} \frac{\Gamma}{2\pi r} r^2 dr = \frac{\Gamma}{2\pi}\int_{r_0}^{R} r dr = \frac{\Gamma}{2\pi}\frac{1}{2}\left(R^2 - r_0^2\right) = \frac{\Gamma}{4\pi} R^2\left(1 - \xi^2\right).$$

Thus,

$$N = M\omega = 2\pi\rho\omega V_1 \frac{\Gamma}{4\pi} R^2\left(1 - \xi^2\right) = \frac{1}{2}\rho\omega V_1 \Gamma R^2\left(1 - \xi^2\right).$$

Since $N = M\omega = \overline{\beta}\rho n^3 d^5$, where $\overline{\beta}$ — propeller power coefficient, we shall find

$$\overline{\beta}\rho n^3 d^5 = \frac{1}{2}\rho\left(1 - \xi^2\right)\omega V_1 \Gamma R^2.$$

Taking into account that $\omega = 2\pi n$, $V_1 = V_0 + v_1$, $v_1 = \overline{V}\omega R = 2\pi n R\overline{V}$, $R = d/2$ we get

$$\overline{\beta}\rho n^3 d^5 = \frac{1}{2}\rho\left(1 - \xi^2\right) 2\pi n\left(V_0 + 2\pi n\frac{d}{2}\overline{V}\right)\Gamma\frac{d^2}{4}.$$

After factoring $\rho n d^2$ out, we get

$$\overline{\beta} n^2 d^3 = \frac{\pi}{4}\left(1 - \xi^2\right)\left(V_0 + \pi n d\overline{V}\right)\Gamma.$$

Let us introduce the dimensionless circulation $\overline{\Gamma} = \frac{\Gamma}{\omega R d}$, then

$$\overline{\beta} n^2 d^3 = \frac{\pi}{4}\left(1 - \xi^2\right)\left(V_0 + \pi n d\overline{V}\right)\overline{\Gamma} 2\pi n\frac{d^2}{2}.$$

Dividing by nd^2, we get

$$\overline{\beta} n d = \frac{\pi^2}{4}\left(1 - \xi^2\right)\left(V_0 + \pi n d\overline{V}\right)\overline{\Gamma}.$$

It follows that

$$\overline{\beta} = \frac{\pi^2}{4}\left(1-\xi^2\right)\left(\frac{V_0}{nd}+\pi\overline{V}\right)\overline{\Gamma}, \quad \text{or}$$

$$\overline{\beta} = \frac{\pi^2}{4}\left(1-\xi^2\right)\left(\lambda+\pi\overline{V}\right)\overline{\Gamma}.$$

Consequently,

$$\overline{\Gamma} = \frac{4\overline{\beta}}{\pi^2\left(1-\xi^2\right)\left(\lambda+\pi\overline{V}\right)}, \quad \text{or}$$

$$\overline{\Gamma} = \frac{4\overline{\beta}}{\pi^3\left(1-\xi^2\right)\left(\frac{\lambda}{\pi}+\overline{V}\right)}.$$

Using the expression obtained for $\overline{V}$, we find

$$\frac{\lambda}{\pi}+\overline{V} = \frac{\lambda}{\pi}-\frac{\lambda}{2\pi}+\sqrt{\frac{\lambda^2}{4\pi^2}+\frac{2\overline{\alpha}}{\pi^3\left(1-\xi^2\right)}} = \frac{\lambda}{2\pi}+\sqrt{\frac{\lambda^2}{4\pi^2}+\frac{2\overline{\alpha}}{\pi^3\left(1-\xi^2\right)}}.$$

And finally,

$$\overline{\Gamma} = \frac{4\overline{\beta}}{\pi^3\left(1-\xi^2\right)\left[\frac{\lambda}{2\pi}+\sqrt{\frac{\lambda^2}{4\pi^2}+\frac{2\overline{\alpha}}{\pi^3\left(1-\xi^2\right)}}\right]}. \tag{5.6}$$

If the propeller operating mode — λ, $\overline{\alpha}$, $\overline{\beta}$ — is specified and the dimensionless hub diameter is known ξ, formula (5.6) enables the determination of the strength of the axial vortex generated by the propeller.

By introducing the dimensionless circulation of the axial vortex through formulas adopted for the aircraft as a whole, $\Gamma^* = \frac{\Gamma}{V_0 L}$, where L — a certain characteristic length, $\overline{\Gamma}$ and Γ^* will be tied with the relation $\overline{\Gamma}\omega R d = \Gamma^* V_0 L$, i.e.,

$$\Gamma^* = \overline{\Gamma}\frac{\omega R d}{V_0 L} = \overline{\Gamma}\frac{2\pi n d}{V_0}\frac{d}{2L};$$

$$\Gamma^* = \overline{\Gamma}\frac{\pi}{\lambda}\overline{d},$$

where $\overline{d} = d/L$ — relative propeller diameter.

The vortical representation of the propeller stream is shown in Fig. 5.2: n vortices evenly distributed over the circumference with a diameter equal to the propeller diameter model the surface of the propeller slipstream.

The circulation of each of these vortices is Γ^*/n, and their sense of rotation is opposite to that of the axial vortex. In this case, the direction of rotation of the air flow caused by the axial vortex coincides with that of the propeller.

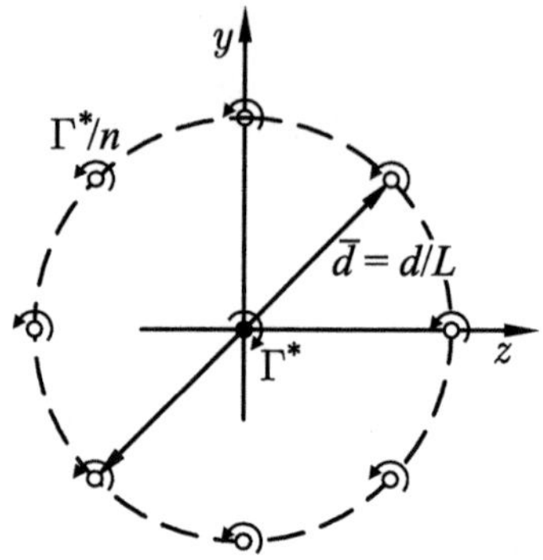

Fig. 5.2. Vortical representation of the propeller slipstream

Thus, formula (5.6) is the basic one for computation of the axial vortex circulation. The question of how many Π-shaped vortices should be taken to accurately simulate the propeller's operation is answered in the following section.

For specifying the operating regime of the propeller, its standard performance data can be used to retrieve the following parameters: $\overline{\beta}$ — propeller power coefficient; λ — propeller speed coefficient, $\overline{\alpha}$ — propeller thrust coefficient, and η — propeller efficiency.

For taking into account the axial velocity in the propeller slipstream, the following approach is used. It is known that the engine thrust P can be obtained with the formula

$$P = G_a(C_j - V),$$

where G_a — air flow rate through the engine, C_j — slipstream velocity, V — flight speed. Having specified the engine power setting or taking the thrust equal to the aircraft's drag, the axial velocity C_j is easily obtainable. Then we determine the ratio between the axial velocity and the flight speed. From the triangle of velocities constructed, we obtain the location of the axial vortex in the next control cross-flow section.

5.3. On a rational number of vortices for modeling a propeller

For determining the number n of Π-shaped vortices required to model a propeller (Fig. 5.3), the following methodological study has been performed. The axial vortices from propellers were calculated for the An-26 and An-12 aircraft under the same flight conditions, but for different numbers of Π-shaped vortices. In so doing, the convergence of the position of the vortices' track in the cross-flow section corresponding to a specified instant of time elapsed after aircraft passage through that section was considered. The vortices' track was assumed "true"at $n \to \infty$. The computed results one presented in Figs. 5.3. and 5.4.

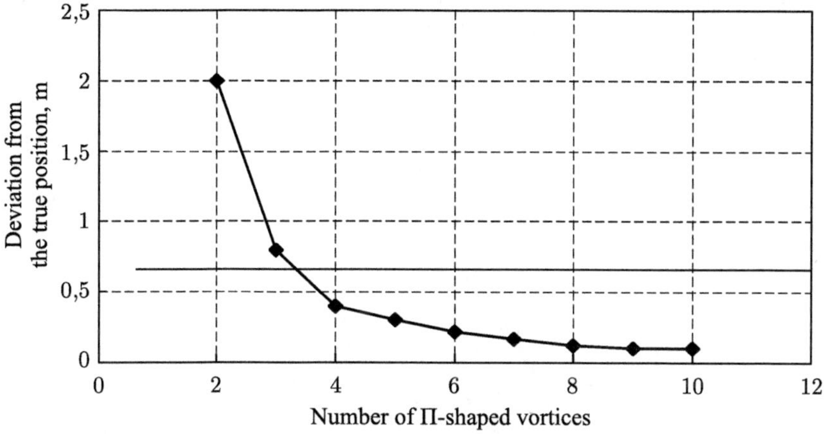

Fig. 5.3. On the determination of a rational number of vortices to model propeller slipstream for the An-26 aircraft

Fig. 5.3 shows the accuracy of computation of the axial vortex modeling the propeller slipstream of the An-26 aircraft flying at a height of 500 m and speed of 450 km/h. It can be seen that at $n > 4$ the vortex' track deviation from the "true"location does not exceed 0,5 m. Similar results are presented in Fig. 5.4 for the An-12 aircraft flying at the same height and speed. It is seen that in this case the indicated deviation does not exceed 0,5 m at $n > 4$ either. Thus it has been found that for modeling the propeller's slipstream at least four П-shaped vortices are required.

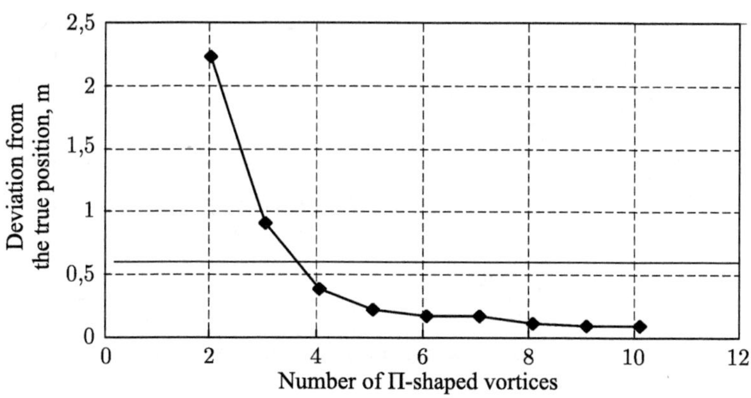

Fig. 5.4. On the determination of a rational number of vortices to model propeller slipstream for the An-12 aircraft

5.4. Examples of computed far vortex wake characteristics of propeller-driven aircraft in comparison with experimental data

For demonstrating the credibility of the above methodology and the reliability of data obtainable with it for propeller-driven aircraft, computations were performed of the vortex wake behind the C-130 aircraft and compared with flight experiments. Fig. 5.5 demonstrates the distribution of the vertical velocity in the vortex core at stations $X = 0$ and $X = 1{,}4$ km behind the C-130 aircraft flying at a height of 1,000 m

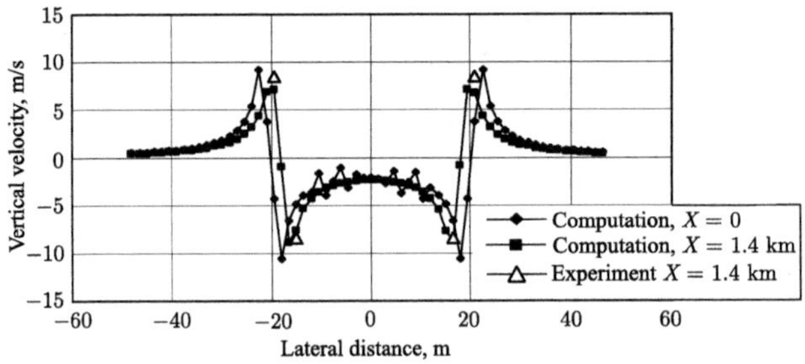

Fig. 5.5. Vertical velocity distribution in the C-130 vortex core

(diamonds and squares — computation for $X = 0$ and $X = 1{,}4$ km, respectively; triangles — experiment, $X = 1{,}4$ km [72].) One can see a satisfactory agreement between theory and experiment.

Shown in Fig. 5.6 is the descent at the C-130 aircraft vortex tubes, computed with the above method and obtained in a flight experiment

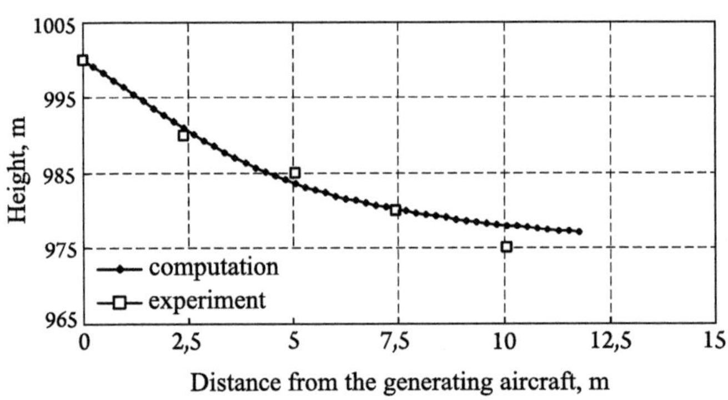

Fig. 5.6. Descent of the vortex tubes behind the C-130 aircraft

[72]. Here one can also see a satisfactory agreement between theory and experiment. These data can serve as a confirmation of the credibility of

the methodology developed and the reliability of computational results obtainable with this methodology.

5.5. The characteristics of the vortex wake behind the An-26 aircraft

At the first stage of studying the characteristics of the vortex wake behind the An-26 aircraft, the following flight regimes were considered:

— flight height 500 m, flight speed 420 km/h, CN= 2;

— flight height 5,000 m, flight speed 420 km/h, CN= 2.

The selection of these two regimes was made from practical considerations.

Fig. 5.7 demonstrates the distribution of the vertical velocity in the vortex core at distances $X = 10$ m, 250 m and 500 m behind the An-26 aircraft flying at $H = 500$ m and $V = 420$ km/h,. It can be seen that propellers noticeably affect the characteristics of the vortex wake at $X = 10$ m. With increasing X, the effect of the propellers decreases and at greater distances disappears completely. The velocities in the vortex core at $X = 0$ m reach 6 m/s in magnitude. As X increases, the vertical velocity in the vortex core diminishes.

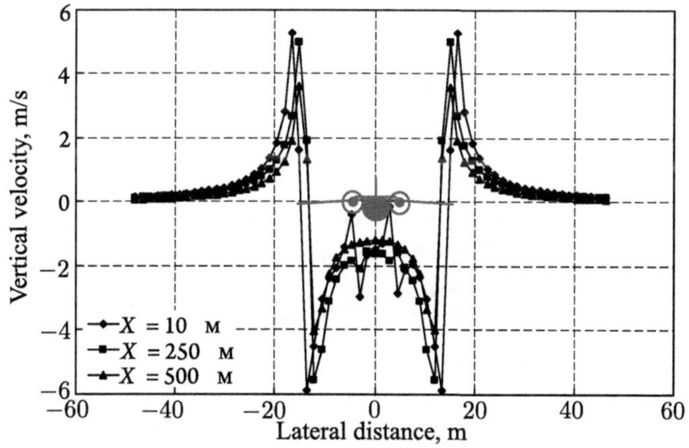

Fig. 5.7. Vertical velocity distribution in the vortex core of the An-26 aircraft

For the An-26 aircraft flying at $H = 5{,}000$ m and $V = 420$ km/h,, the vertical velocity distributions are given in Fig. 5.8 at distances of $X = 10$ m, 250 m and 500 m downstream of the aircraft. In this case the propellers significantly affect the vortex wake characteristics only at $X = 10$ m. As X increases, the influence at the propellers diminishes and later also disappears completely. The velocities in the vortex core at $X = 10$ in this case reach 8 m/s in magnitude. At greater distances the vertical velocity is the vortex core decreases.

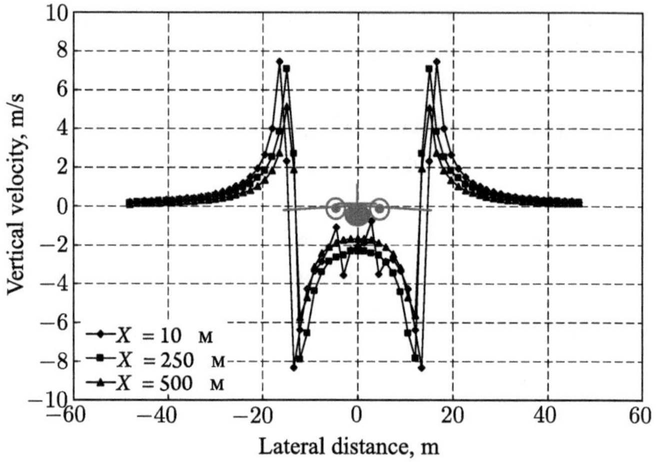

Fig. 5.8. Vertical velocity distribution in the vortex core of the An-26 aircraft

The second stage of the study was devoted to investigation of the far vortex wake at a distance of 12,5 km behind the An-26. The results of this study are as follows.

Fig. 5.9. depicts the descent of the vortex wake behind the An-26 aircraft for various atmospheric conditions on the adopted 5-point scale. It is seen that in a stable atmosphere the wake's loss of height is the greatest.

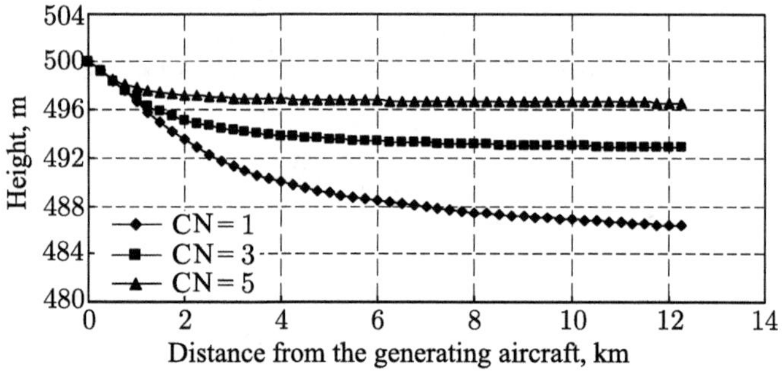

Fig. 5.9. Descent of the vortex wake behind the An-26 aircraft

The reason is that in such an atmosphere the intensity (circulation) of the wing-tip vortex persists longer, which is confirmed by Fig. 5.10, presenting a time dependence of the circulation (m^2/s) of a wing-tip vortex. In a very unstable atmosphere (CN= 5) the wing-tip vortices decay more rapidly. These facts are known from previous experimental and flight investigations [8, 14–18] and the findings of this numerical study support the earlier data.

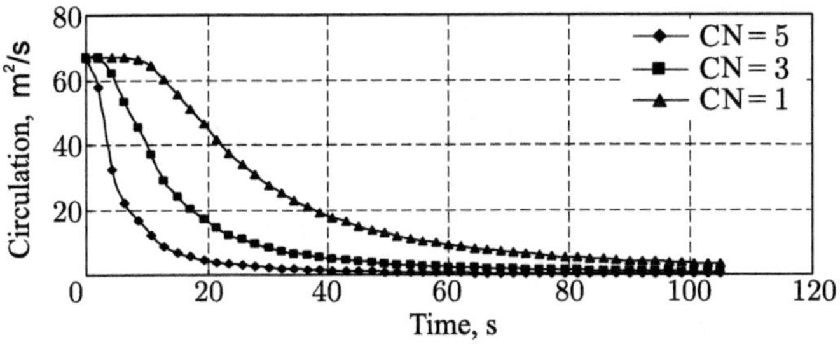

Fig. 5.10. Time history of the circulation of the An-26's wing-tip vortex

Shown in Fig. 5.11 is the vertical velocity distribution in the vortex core at distances of $X = 10$ m, 1,000 m, 5,000 m and 12,500 m behind the An-26 aircraft flying at $H = 5{,}000$ m and $V = 420$ km/h, for CN= 1. In this case, with distance X from the aircraft increasing the vertical

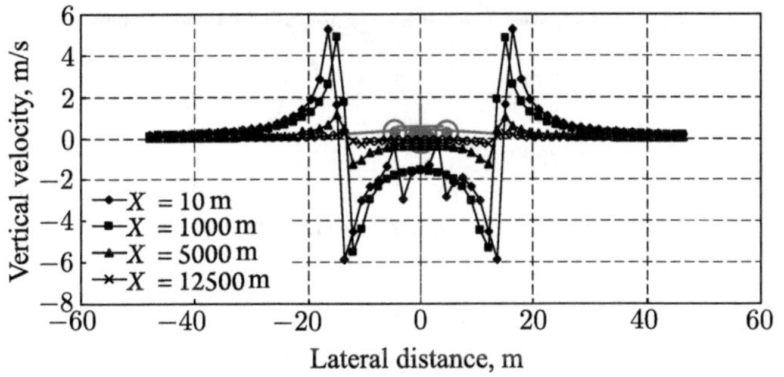

Fig. 5.11. Vertical velocity distribution in the vortex core of the An-26 aircraft, CN= 1

velocities in the wing-tip vortex core decrease. At $X = 1{,}000$ m, they are still noticeable (5–6 m/s), but at $X = 5{,}000$ m do not exceed 2 m/s. Similar results are presented for other atmospheric conditions. The velocity distributions for the An-26 aircraft are shown in Figs. 5.12 and 5.13 for CN= 3 and CN= 5, respectively. It is seen that in an unstable atmosphere air flow disturbances attenuate more rapidly.

Let us now consider takeoff and landing operations. The extension of the high-lift devices influences the location of the vortices shed from the wing and flap tip (Fig. 5.14). Besides, the strength of the vortices shed from the flap is by a factor of several times greater than that of the wing-tip vortices (Fig. 5.15). As a result of these vortices' interaction and the influence of the propellers' vortices, the symmetry between the locations of right and left vortex systems becomes broken (Fig. 5.14). It is also observed that the wing-tip vortices approach each other (their separation becomes 8–10 m) and loops appear in the tracks of the flap vortices in control planes.

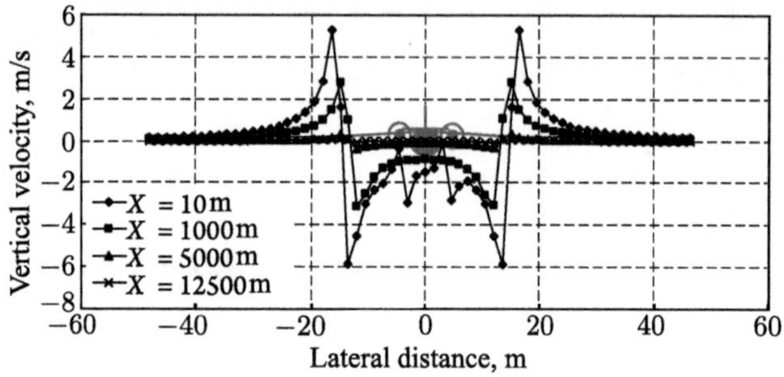

Fig. 5.12. Vertical velocity distribution in the vortex core of the An-26 aircraft, CN= 3.

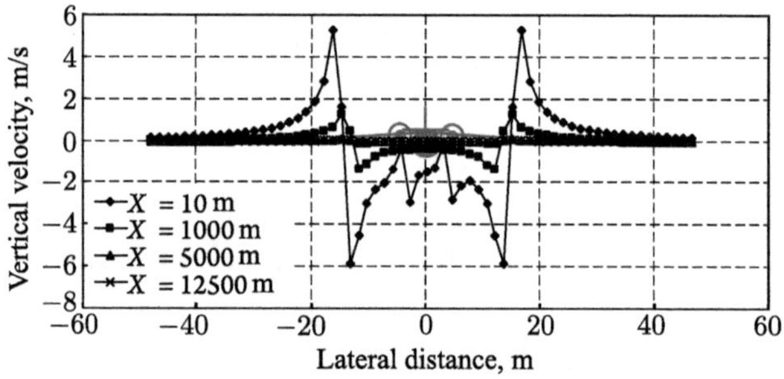

Fig. 5.13. Vertical velocity distribution in the vortex core of the An-26 aircraft, CN= 5

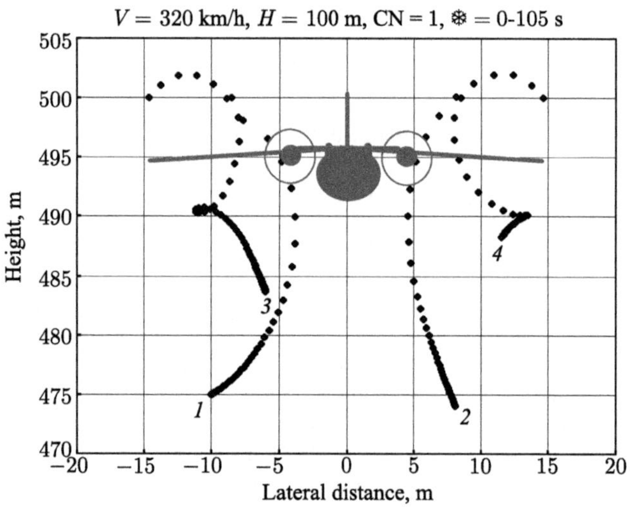

Fig. 5.14. Tracks of the wing-tip vortices (*1, 2*) and the flap vortices (*3, 4*) in a control cross-flow plane behind the An-26 aircraft.

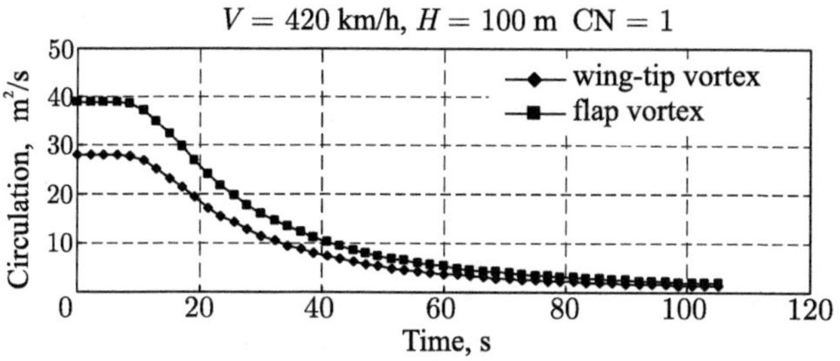

Fig. 5.15. Time history of the circulation of the An-26's vortices

With decreasing flight height and speed the intensity of the vortices increases (Fig. 5.16). This, in turn, leads to changes in vortex shapes (Fig. 5.17). Fig. 5.17 demonstrates the wing-tip and flap vortices behind the aircraft at takeoff power at $V = 180$ km/h, $H = 50$ m, CN$= 1$ within the time period of $T = 0–150$ s at $\Delta T = 1$ s intervals. In this case the propellers do not affect significantly vortex configurations. With engines operating at idle, the situation becomes different.

Fig. 5.18 shows the vortices behind the An-26 aircraft flying at $H = 25$ m and $V = 180$ km/h. The time period considered is $T = 0–150$ s. The points showing vortex position are presented at $\Delta T = 1$ s time intervals.

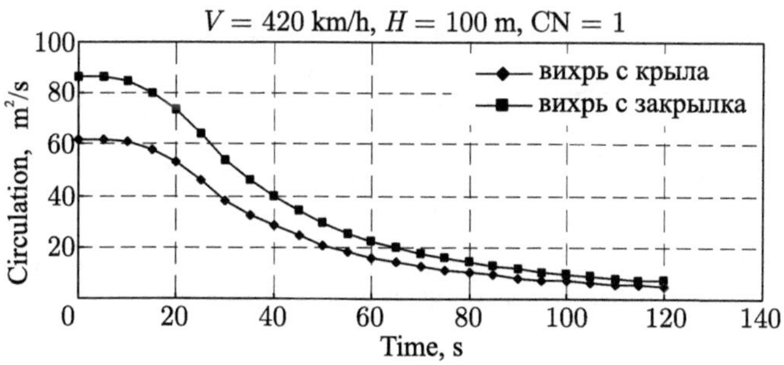

Fig. 5.16. Vortex circulation time history for the An-26 aircraft

Computations revealed a rising of the wing-tip vortices as a result of the high-lift devices' extension (Figs. 5.14, 5.17, 5.18). The reason is the interaction of the wing and flap vortices whose strengths for the An-26 at landing are of the same order of magnitude (see Fig. 5.16).

The methodology developed allows the calculation of the vortex wake behind propeller-driven aircraft at various flight conditions.

Thus, the studies of the vortex wake behind the An-26 aircraft have shown that the propellers significantly affect the wake's characteristics only at short distances from the aircraft (up to 500 m). At greater distances (up to 12,5 km) the propellers' influence is inessential. During

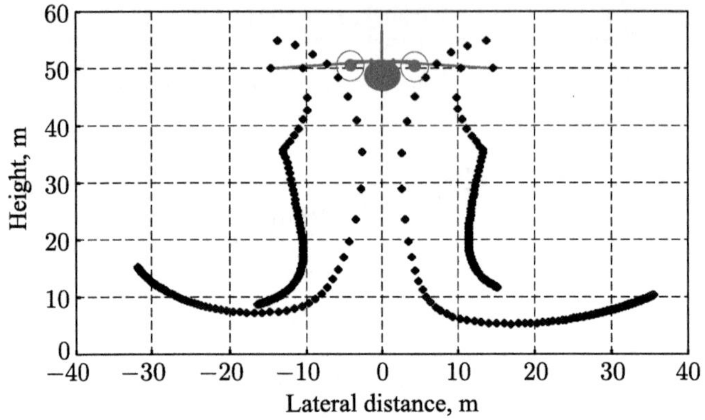

Fig. 5.17. Tracks of wing-tip and flap vortices in a control cross-flow plane behind the An-26 aircraft at takeoff power

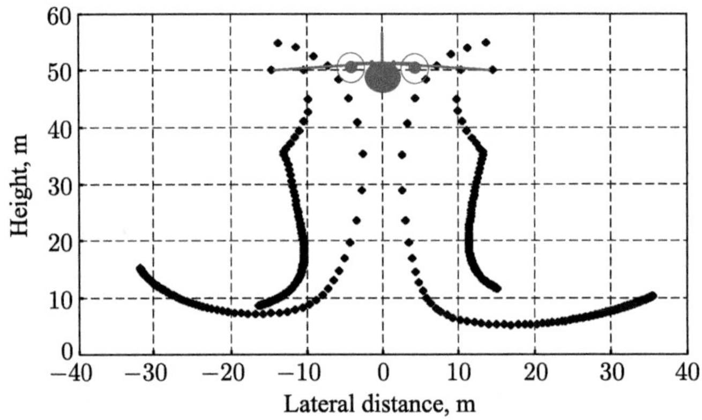

Fig. 5.18. Tracks of the wing-tip and flap vortices in a control cross-flow plane behind the An-26 aircraft at idle power

takeoff and landing operations with high-lift devices deployed the effect of propellers manifests itself through the asymmetry of the left and right vortex systems of the aircraft.

5.6. The characteristics of the vortex wake behind the An-12 aircraft

This section deals with the results of investigations into the characteristics of the far vortex wake behind the An-12 aircraft. These investigations were performed in two phase: the first was devoted to detailed studies of the vortex wake at distances behind the aircraft up to 500 m, the second treated, although somewhat less thoroughly, the wake's characteristics at distances up to 12,5 km.

The following flight regimes were considered in the first phase:
— flight height 500 m, flight speed 420 km/h, CN= 1;
— flight height 5,000 m, flight speed 420 km/h, CN= 1.
These regimes were selected from practical considerations.

Fig. 5.19 demonstrates the vertical velocity distribution in the core of the vortex at distances $X = 10$ m, 150 m, 300 m and 500 m behind the An-12 aircraft flying at $H = 500$ m, $V = 420$ km/h, CN= 1. The effect of propellers was ignored in computation.

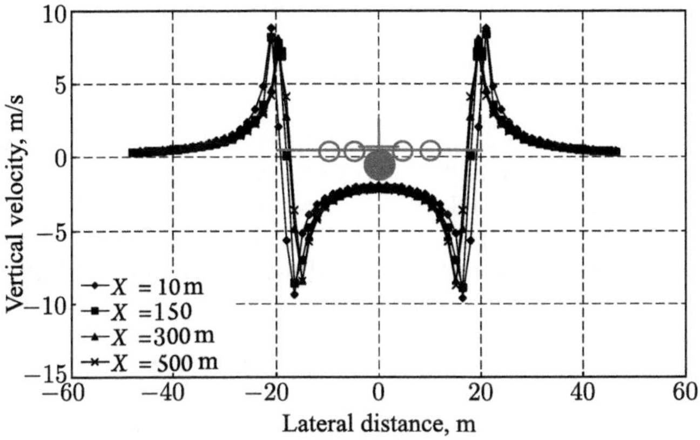

Fig. 5.19. Vertical velocity distribution in the core of a vortex behind the An-12 aircraft (propeller effect ignored).

Similar data are presented in Fig. 5.20 with the propellers' effect taken into account. One can see that the propellers noticeably affect the wake's characteristics at distances up to $X = 150$ m. As X increases, the

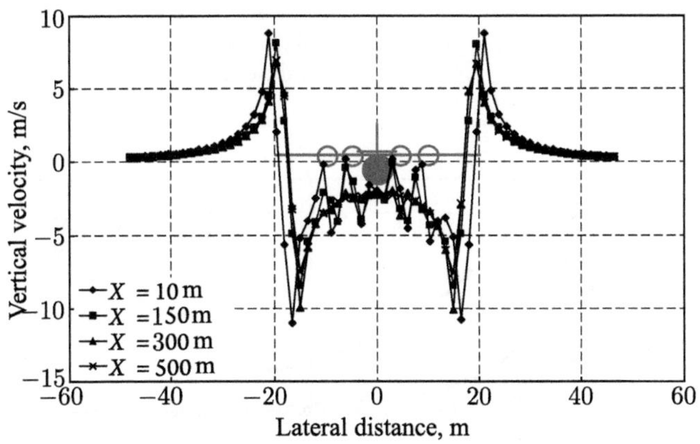

Fig. 5.20. Vertical velocity distribution in the core of a vortex behind the An-12 aircraft (propeller effect taken into account).

influence of the propellers weakens and then disappears completely. In this case, the velocities in the vortex core at $X = 10\,\text{m}$ reach 10–12 m/s in magnitude. With increasing X, he vertical velocity in the vortex core decreases.

In phase 2 studies were devoted to the far vortex wake behind the An-12 at downstream distances up to 12,5 km. These results one presented below.

Fig. 5.21 shows the descent of the vortex wake behind the An-12 aircraft flying at $V = 420\,\text{km/h}$,, $H = 50\,\text{m}$ under various atmospheric conditions. It is seen that in a stable atmosphere the wake descends significantly, to heights of 25–27 m. This is associated with the fact that in a stable atmosphere the strength (circulation) of a wing-tip vortex persists longer. Fig. 5.22 shows the time history of the circulation of the wing-tip vortex behind the An-12 flying at the same height and velocity. In a highly unstable atmosphere the wing-tip vortex decays more rapidly. This is known (as was noted earlier) from laboratory and flight investigations [8, 14–18] and confirmed by the computed results presented here.

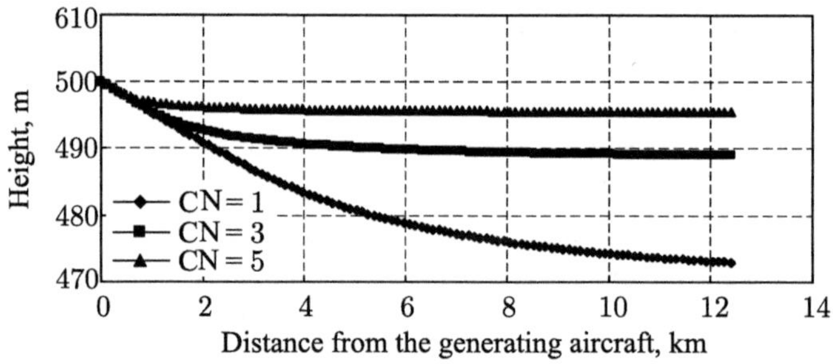

Fig. 5.21. Descent of the vortex wake behind the An-12 aircraft at various atmospheric conditions

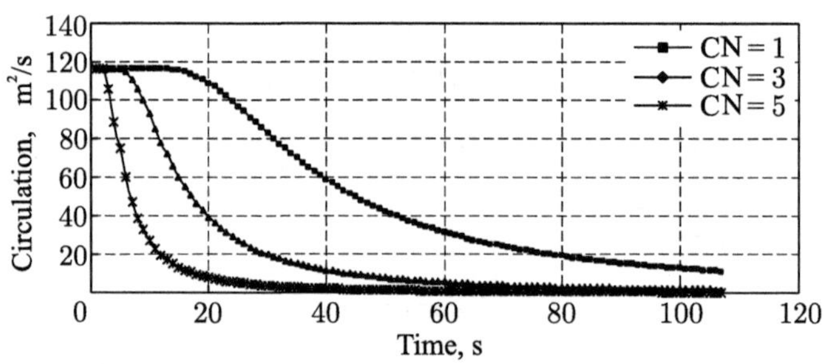

Fig. 5.22. Time history of the circulation of a vortex behind the An-12 aircraft

The effect of the propellers on the descent of the left and right wing-tip vortices of the An-12 with engines at cruise power is shown in Fig. 5.23. One can see that the left vortex descends lower, which is associated with the same directions of rotation of the propellers and the left vortex.

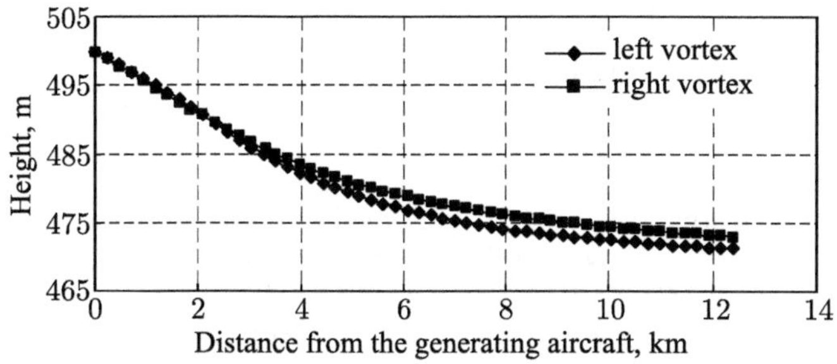

Fig. 5.23. Descent of the left and right wing-tip vortices behind the An-12 aircraft

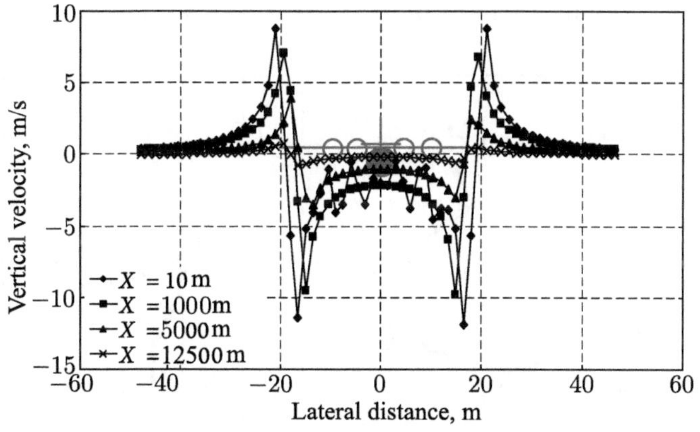

Fig. 5.24. Vertical velocity distribution in the core of a wing-tip vortex behind the An-12 aircraft

Fig. 5.24 depicts the vertical velocity distribution in the vortex core of a vortex at distances $X = 10\,\mathrm{m}$, $1{,}000\,\mathrm{m}$, $5{,}000\,\mathrm{m}$ and $12{,}500\,\mathrm{m}$ behind the An-12 flying at $H = 500\,\mathrm{m}$, $V = 420\,\mathrm{km/h}$, CN= 1. With increasing X, the vertical velocities in the core of a wing-tip vortex decreases, but at a distance of $5{,}000\,\mathrm{m}$ they are still 3 – 4 m/s.

Consider takeoff/landing operations. High-lift devices deployed significantly affect the characteristics of the wing-tip and flap vortices (Fig. 5.25). The intensity of vortices leaving the flap is twice the strength of the wing-tip vortices (Fig. 5.26). The interaction of these vortices and the effect of propellers break the symmetry of the left and right vortex systems (Fig. 5.25). Besides, the vortices move closer to one another to distances of 10 – 12 m and loops appear in the flap vortices' tracks in the control cross-flow plane.

With decreasing flight height and speed, the vortices' intensity increases. (Fig. 5.27). This, in turn, leads to changes in their geometry (Fig. 5.28). Fig. 5.28 depicts the tracks of the vortices from the wing-tips and flaps in the control section for the aircraft flying at $V = 180\,\text{km/h}$,, $H = 50\,\text{m}$, CN= 1 within the time period $T = 0 - 150\,\text{s}$. The points in the figure are separated by $\Delta T = 1\,\text{s}$ intervals. The engines operate at cruise power setting, and in this case the propellers significantly affect the vortices.

When selecting the takeoff power setting, the situation changes still more radically, as shown in Fig. 5.29. The An-12 passed through the control plane at $H = 50\,\text{m}$ and $V = 180\,\text{km/h}$. The series of the vortex tracks covers the time period $T = 0 - 150\,\text{s}$. Neighboring points are separated by $\Delta T = 1\,\text{s}$ intervals.

The computations reveal a rising of the An-12 wing-tip vortices with extending the high-lift devices (Figs. 5.25, 5.28, 5.29). This is caused by the interaction between the wing-tip and flap vortices, whose intensities for the An-12 differ more than by a factor of two (Figs. 5.26 and 5.27).

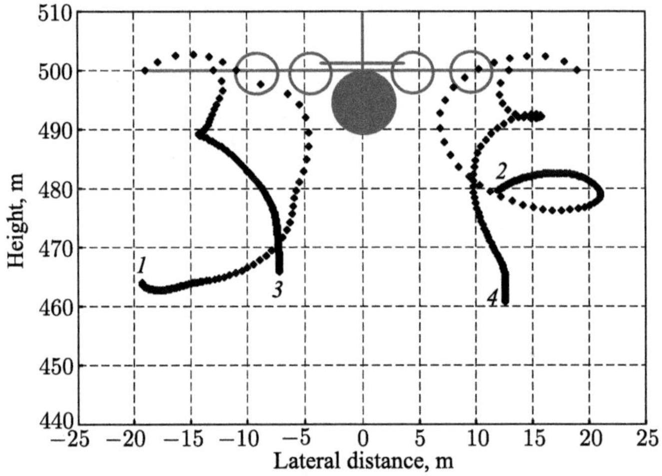

Fig. 5.25. Tracks of the wing-tip (*1, 2*) and flap (*3, 4*) vortices of the An-12 aircraft in a control cross-flow plane

The methodology developed allows one to compute the vortex wake behind an aircraft with four propellers for other flight conditions as well.

Thus, the studies into the vortex wake of the An-12 aircraft have shown a noticeable effect of the propellers on the vortex wake characteristics both in cruise and takeoff/landing flight regimes. The propellers' influence manifests itself through the asymmetry of the right and left

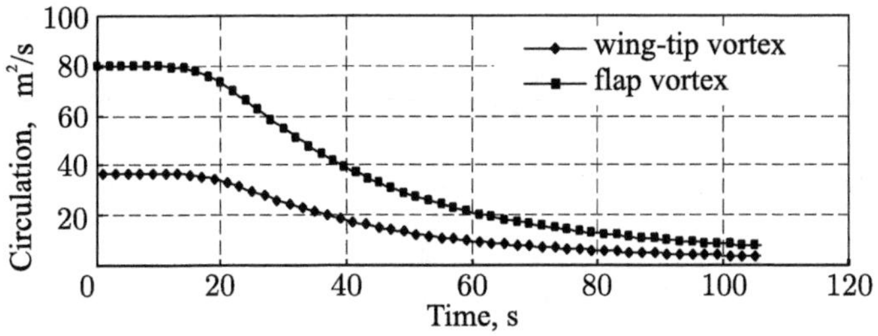

Fig. 5.26. Time history of the vortex circulation for the An-12 aircraft flying at ($H = 500\,\text{m}$, $V = 180\,\text{km/h}$)

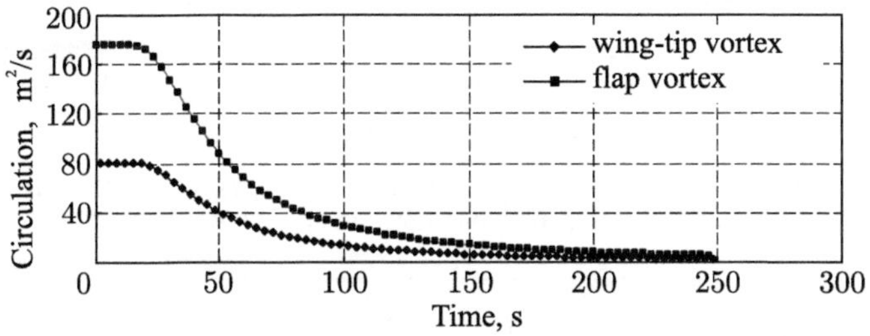

Fig. 5.27. Time history of the vortex circulation for the An-12 aircraft flying at $H = 50\,\text{m}$ $V = 180\,\text{km/h}$.

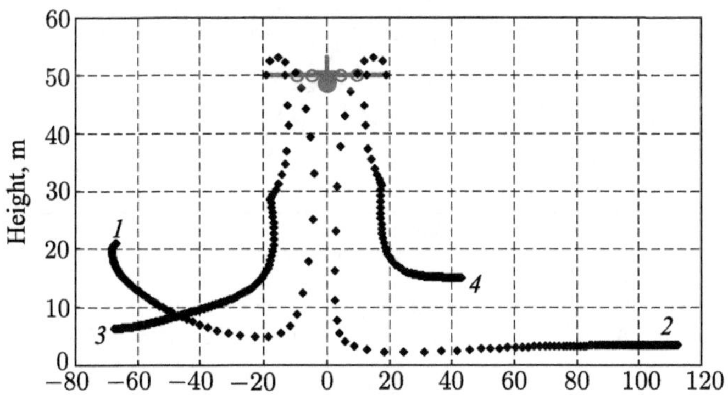

Fig. 5.28. Tracks of the wing-tip (*1, 2*) and flap (*3, 4*)) vortices of the An-12 aircraft and a control cross-flow plane (cruise regime)

vortex systems of an aircraft. Especially strong is the effect of propellers on the vortex wake during takeoff and landing operations with high-lift devices deployed.

Fig. 5.30 depicts the vertical velocity distribution in the core of a wing-tip vortex of the An-12 aircraft at takeoff-power setting, $V = 220\,\text{km/h}$,, $H = 50\,\text{m}$, CN= 1. It is seen that the high-lift devices deployed and the

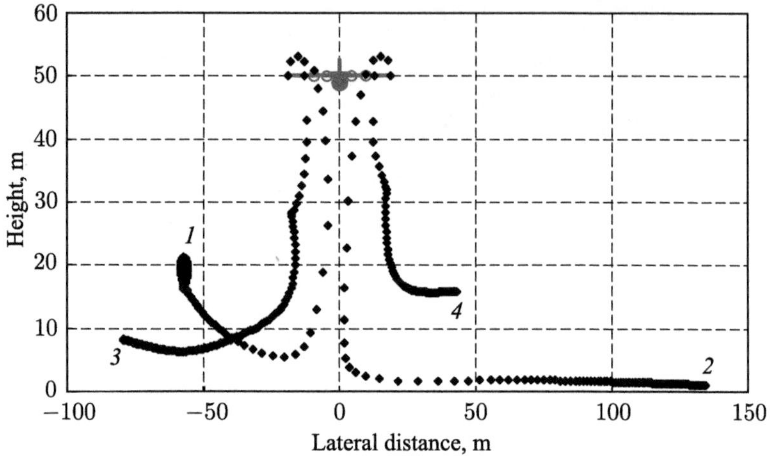

Fig. 5.29. Tracks of the wing-tip (*1, 2*) and flap (*3, 4*)) vortices of the An-12 aircraft in a control cross-flow plane (takeoff regime)

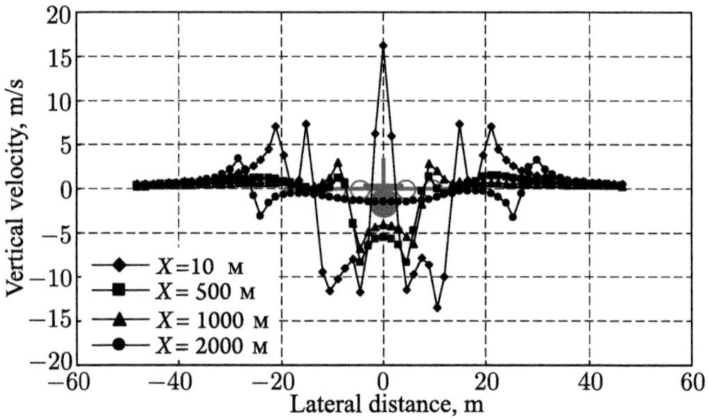

Fig. 5.30. Vertical velocity distribution in the core of a wing-tip vortex of the An-12 aircraft with its high-lift devices deployed

propellers running result in noticeable oscillations of the vertical velocity with the z coordinate.

5.7. The characteristics of the vortex wake behind the C-130 aircraft

The studies into the characteristics of the vortex wake of the C-130 aircraft is of interest from the standpoint of the comparison with the An-12. The takeoff weights of the C-130 and An-12 are 80 t and 60 t, respectively. They have about identical wing spans: 40 m for the C-130

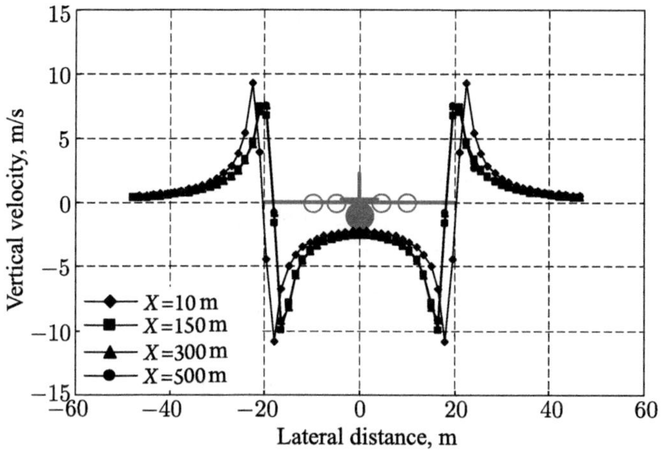

Fig. 5.31. Vertical velocity distribution in the core of a wing-tip vortex behind the C-130 aircraft (no propeller effect taken into account)

and 39 m for the An-12. As previously for the An-12, the following flight regimes were considered for the C-130:

— flight height 500 m, flight speed 420 km/h, CN= 1;

— flight height 5,000 m, flight speed 420 km/h, CN= 1.

These regimes were selected from practical considerations.

Fig. 5.31 demonstrates the vertical velocities in the core of a wing-tip vortex at distances $X = 10$ m, 150 m, 300 m and 500 m behind the C-130 aircraft flying at $H = 500$ m and $V = 420$ km/h. Here, the effect of the propellers on the vortex wake's characteristics was not taken into consideration.

Similar computed data are presented in Fig. 5.32, but now with account for the propellers' effect. It can be seen that the propellers significantly affect the vortex wake's characteristics at distances up to $X = 150$ m. As X increases, the effect of the propeller weakens and disappears completely at greater distances. The velocities in the vortex core at $X = 10$ m reach 10 – 12 m/s in magnitude. At greater X, the vertical velocities in the vortex core decrease together with the effect of the propellers.

Shown in Fig. 5.33 is the vertical velocity distribution in the core of the wing-tip vortex of the C-130 at distances $X = 10$ m, 250 m and 500 m behind the aircraft flying at $H = 5{,}000$ m, $V = 420$ km/h,. It is seen that near the aircraft the propeller significantly affect the vortex wake's parameters. As X increases, the propellers' effect weakens and disappears completely at greater distances. The velocities in the core at $X = 10$ increase, reaching 14 – 16 m/s in magnitude, but at greater distances the vertical velocity in the core decreases.

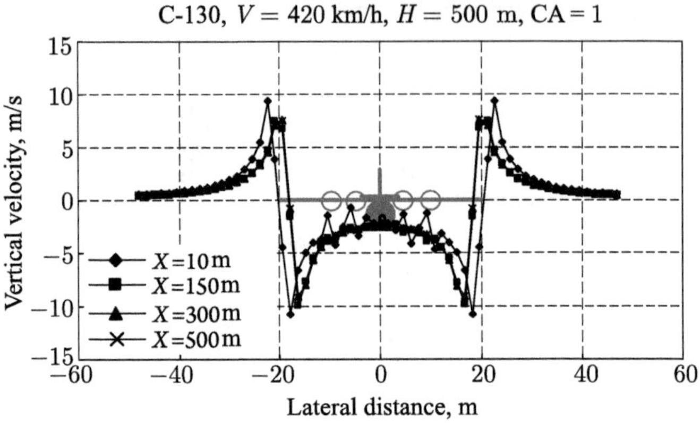

Fig. 5.32. Vertical velocity distribution in the core of a wing-tip vortex of the C-130 aircraft (with account for propeller effect)

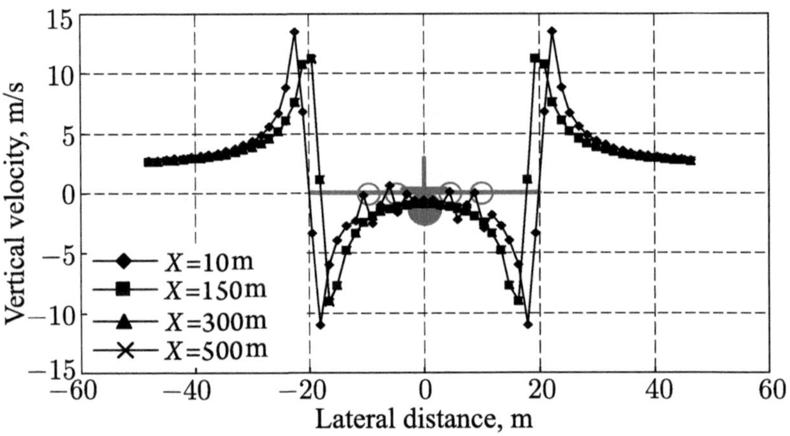

Fig. 5.33. Vertical velocity distribution in the core of a wing-tip vortex of the C-130, $H = 500$ м (with propellers' effect taken into consideration)

The C-130 far vortex wake was investigated at distances up to 12.5 km. Fig. 5.34 demonstrates the descent of the C-130 vortex wake at various atmospheric conditions: CN= 1, 3 and 5. It can be seen that in a stable atmosphere the wake sinks most significantly, about 40 m. This is due to the fact that under such conditions the strength (circulation) of the wing-tip vortex persists longer. Fig. 5.35 shows the time history for the vortex circulation. In a highly unstable atmosphere aircraft wing-tip vortices decay more rapidly.

Shown in Fig. 5.36 are vertical velocity distributions in the core of a wing-tip vortex at distances $X = 10$ m, 1,000 m, 5,000 m and 12,500 m behind the C-130 flying at $H = 5{,}000$ m, $V = 420$ km/h, CN= 1. At $X =$

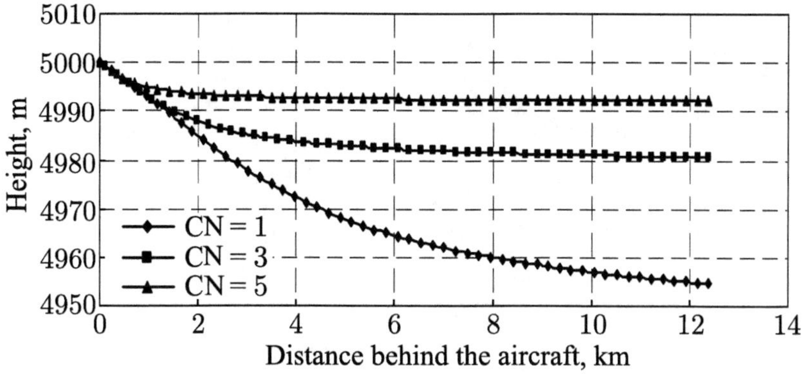

Fig. 5.34. Descent of the C-130 vortex wake at different atmospheric conditions

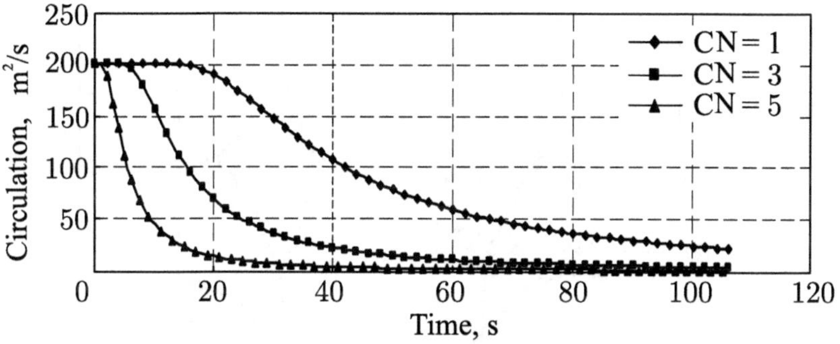

Fig. 5.35. Time history of the wing-tip vortex circulation for the C-130 aircraft

$= 10$ the vertical velocities reach 13 – 15 m/s. As X increases, the vertical velocities in the vortex core decreases. At $X = 1{,}000$ m they are still noticeable (11 – 13 m/s), at $X = 5{,}000$ m they do not exceed 5 – 6 m/s and 2 m/s at $X = 12{,}500$ m.

The deployment of high-lift devices significantly affects the vortices shed from the C-130 wing and flap (Fig. 5.37). These vortices differ in strength by a factor of about 2.5 (Fig. 5.38). The engines even at cruise power setting do not affect noticeably the motion of the vortices. Fig. 5.37 demonstrates the tracks the vortices in the control plane for the aircraft C-130 flying at $H = 100$ m and $V = 320$ km/h,. With setting the engines at takeoff power the situation changes. As a result of interaction of the vortices shed from the wing tips and flaps, as well as due to influence of the propellers, the symmetry of the right and left vortex systems is broken, as shown in Fig. 5.39 for the C-130 flying at $H = 100$ m, $V = 320$ km/h, with its engine at takeoff power. A loop-shaped tracks of the vortices from the wing tip and flap in the control plane is observed.

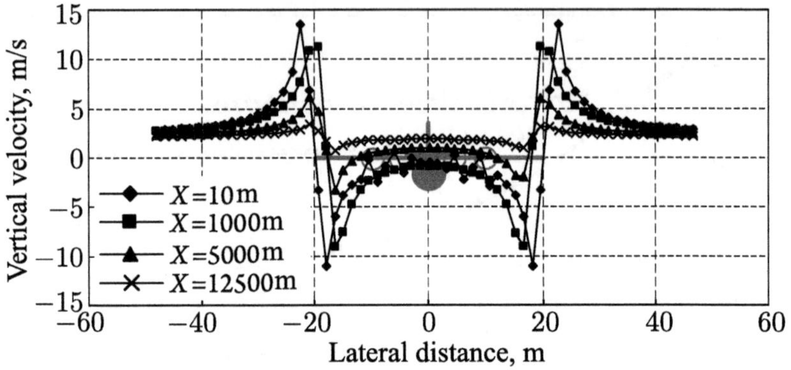

Fig. 5.36. Vertical velocity distribution in the core of a wind-tip vortex of the C-130 aircraft

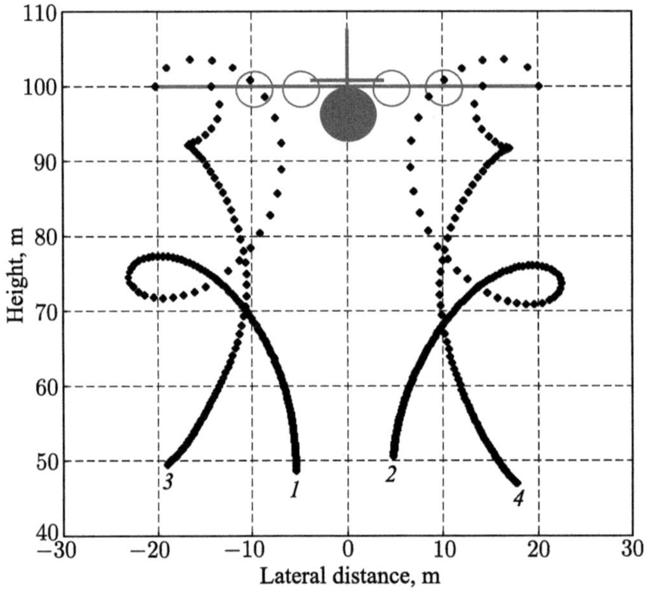

Fig. 5.37. Tracks of the wing-top (*1, 2*) and flap (*3, 4*)) vortices in a control cross-flow plane behind the C-130 aircraft

The vortices' dynamics was observed for the period of time 0 — 150 s. The points on the graph were plotted at a $\Delta T = 1$ s interval. Fig. 5.40 presents the vortex wake behind the C-130 for the flight conditions identical to those used for the previous figure. (Fig. 5.39)

With decreasing flight height and velocity the vortices' strength increases. This, in turn, leads to changes in the wake's configuration. Fig. 5.41 characterizes the vortices shed from the wing and flap of the

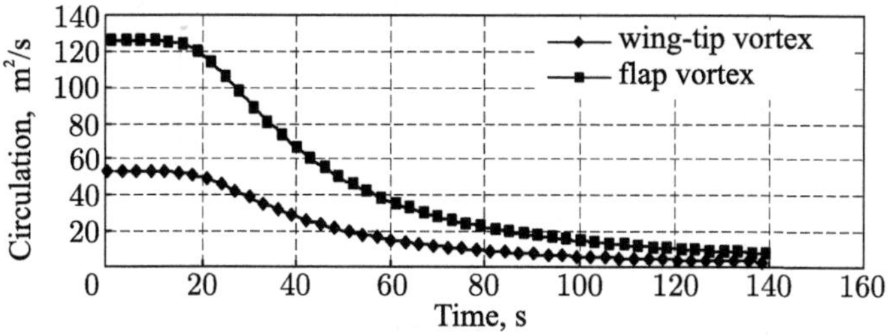

Fig. 5.38. Time history of the circulation of the C-130 vortices (cruise power setting)

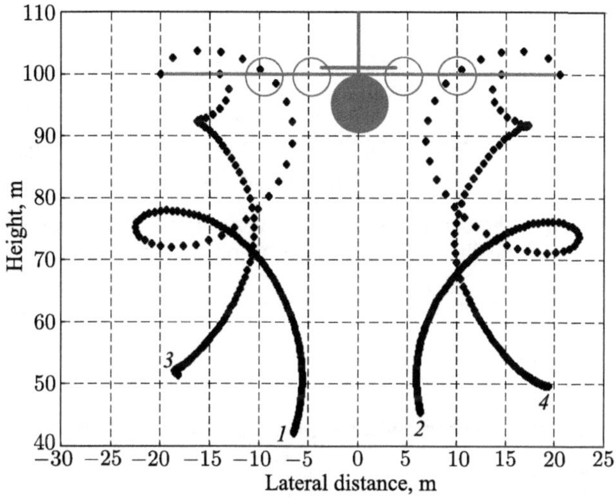

Fig. 5.39. Tracks of the wing-tip (*1, 2*)) and flap (*3, 4*)vortices in a control cross-flow plane behind the C-130 aircraft (takeoff power setting)

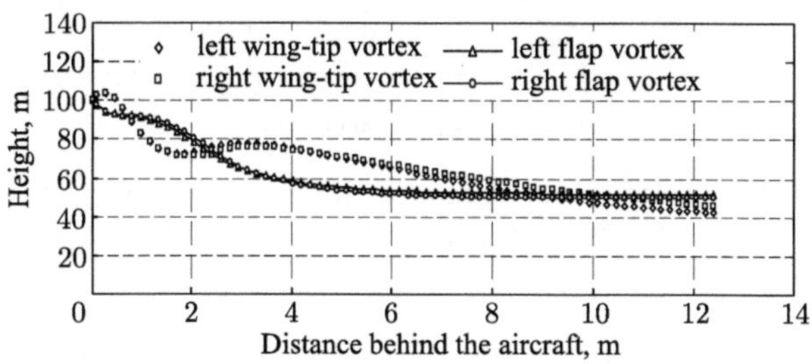

Fig. 5.40. Descent of the vortices of the C-130 aircraft

C-130 flying at $V = 180\,\mathrm{km/h}$, $H = 50\,\mathrm{m}$, CN$= 1$ over the time period of $T = 0 - 150\,\mathrm{s}$. The plotted neighboring points are separated by the $\Delta T = 1\,\mathrm{s}$ time interval. These data are for takeoff power setting. One

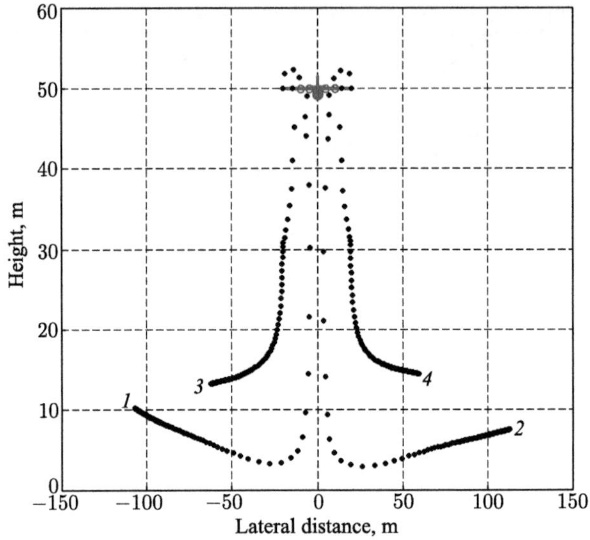

Fig. 5.41. Tracks in the control plane of the wing-tip (*1, 2*) and flap (*3, 4*) vortices behind the C-130 aircraft (takeoff power setting)

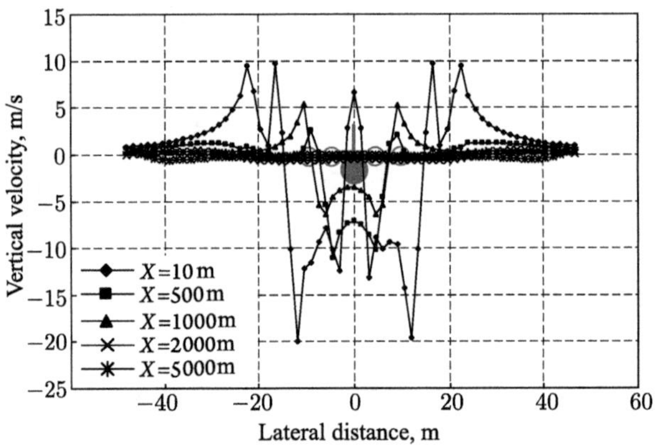

Fig. 5.42. Vertical velocity distribution in the core of a wing-tip vortex of the C-130 aircraft (takeoff power setting)

can see in the control plane that the wing-tip vortices first rise and then begin to descend and approach each other. When approaching the ground surface, they move along it and even rise. The vortices shed from the flaps move apart as a result of mutual interaction between all the vortices. It is also seen that the propellers only weakly affect the motion of the wing-tip and flap vortices of the C-130, as opposed to the case of the An-12 aircraft (see Fig. 5.29). Thus, the investigations of the vortex wake of the C-130 have shown that propellers significantly affect the wake's characteristics only not far from the aircraft (at distances less than 500 m). At greater distances (up to 12,5 km) the influence of propellers is inessential. During takeoff/landing operations with the high-left devices deployed, the effect of propellers manifest itself through the asymmetry of the right and left vortical systems of the aircraft.

Depicted in Fig. 5.42 is the vertical velocity distribution in the core of the vortices behind the C-130 aircraft at $V = 180$ km/h, $H = 50$ m, CN= 1. It is seen that the vertical velocities in the core of the vortices behind the C-130 with the high-lift devices deployed and engines operating reach 20 m/s at a downstream distance of 10 m, 10-12 m/s at $X = 500$ m and 5 – 7 m/s at $X = 1{,}000$ m.

Chapter 6

WIND FLOW OVER ROUGH TERRAIN

6.1. Basic conditions

When flying at low heights, especially during takeoff and landing operations, aircraft can encounter with wakes caused by orographic influence of a terrain (mountains, hills, etc.). This chapter will show how in the framework of the DVM one can model terrain-influenced wind flows. Similar problems arise when studying aircraft operations from aircraft carriers. This issue is investigated in detail in Ref. [8] and is not touched upon in the present monograph.

6.2. Problem statement

Consider air flow at a constant velocity V over a rugged terrain (Fig. 6.1). The problem is to determine a velocity and pressure fields at any point of the air space over the terrain. The following conditions are specified and assumptions made: the medium is an ideal incompressible

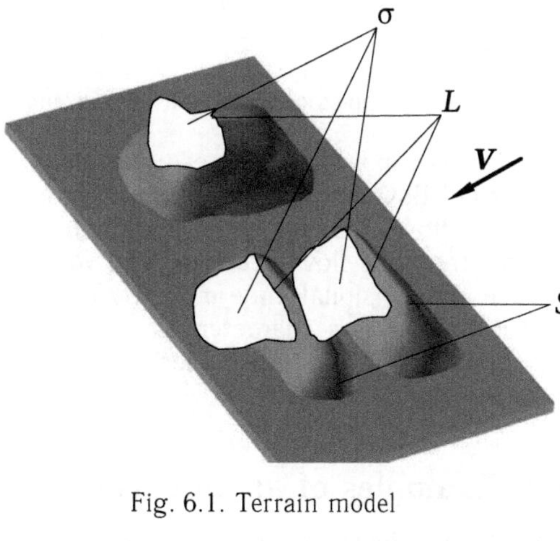

Fig. 6.1. Terrain model

fluid, the flow is potential everywhere except for the surfaces σ and S; the tangency condition is met at the surfaces S, $\left(\vec{\nabla}\varphi, \vec{n}\right) = 0$, $\vec{r_g} \in S$,

A.S. Ginevsky, A.I. Zhelannikov, *Vortex Wakes of Aircrafts*, Foundations of Engineering Mechanics,
DOI 10.1007/978-3-642-01760-5_6, © Springer-Verlag Berlin Heidelberg 2009

no pressure jump across the surfaces σ: $\left(\vec{\nabla}\varphi, \vec{n}\right)_+ = \left(\vec{\nabla}\varphi, \vec{n}\right)_-$, $p_+ = p_-$, $\vec{r_g} \in \sigma$, flow velocities are finite at the edges where vortex sheets leave terrain heights (Chaplygin–Zhukovsky condition); flow disturbances attenuate at infinity, $|\nabla\varphi| \to \vec{V}_\infty$, $|\varphi| \to (\vec{V}_\infty, \vec{r}_g)$, $|\vec{r_g}| \to \infty$. Flow separation lines are specified. The velocity field is determined through the Biot-Savart formula, the pressure field is found using the Cauchy–Lagrange integral, $\dfrac{P}{\rho} = \dfrac{P_\infty}{\rho} - \dfrac{\overline{W}^2_\infty}{2} - \dfrac{\partial U}{\partial t}$.

6.3. A solution technique. Terrain representation

The problem under consideration is solved with the DVM. To do this requires a schematized terrain model (Fig. 6.2) following the surface relief. The model's surface is covered with closed vortex frames. The problem

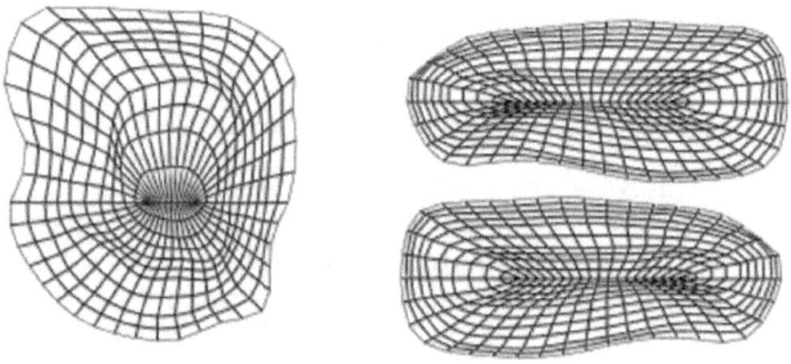

Fig. 6.2. Examples of topography representation

in hand was treated as a nonlinear unsteady problem [4, 7]. An example of the computed wind flow over a terrain is presented in Fig. 6.3, where the vortex sheet is shown leaving three mountains of the same height. An arrow indicates the wind flow direction. The vortex sheet corresponds to an instant of nondimensional time $\tau = Vt/h = 3{,}3$, where V is the wind velocity, t is time, h is a characteristic linear dimension (mountain height). At any point of the airspace under consideration one can obtain the required characteristics of the air flow.

6.4. Examples of air flow computations

Figs. 6.4 and 6.5 demonstrate a disturbance velocity field in the section S at instants of time $\tau = 0{,}15$ and 3,3, respectively. One can see the initiation and development of vortices over a terrain under consideration.

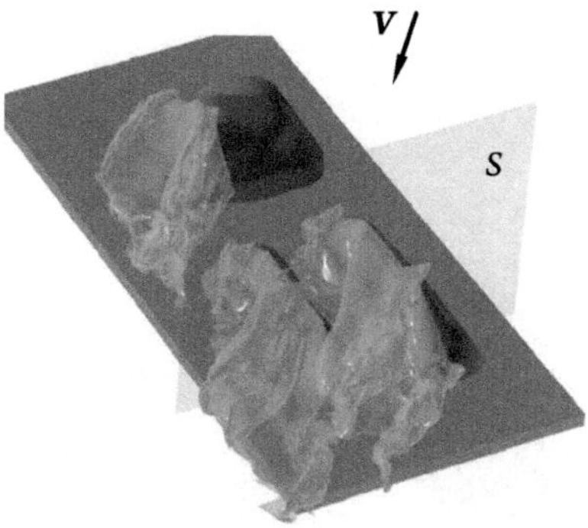

Fig. 6.3. Vortex sheet on the leeward sides of three mountains, $\tau = 3{,}3$

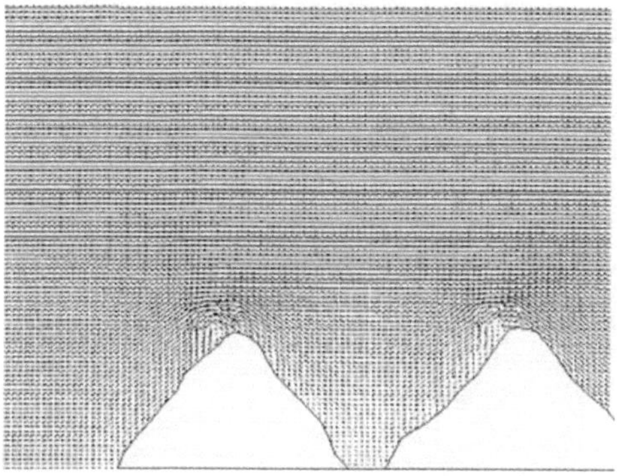

Fig. 6.4. Velocity field in a vertical plane at $\tau = 0{,}15$

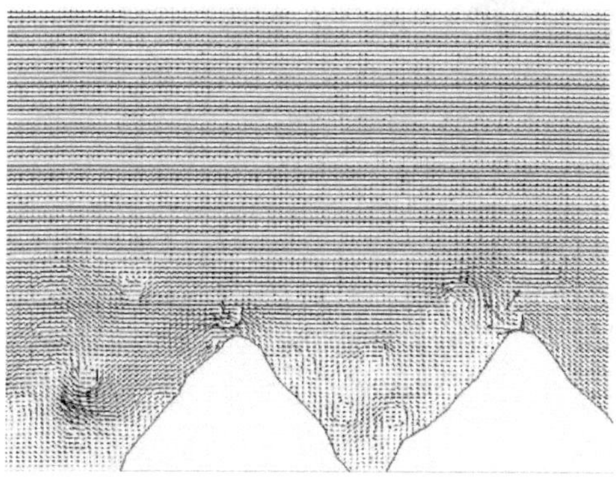

Fig. 6.5. Velocity field in a vertical plane at $\tau = 3{,}3$

Flow angles in the section S at $\tau = 0{,}15$ and 3,3 are presented in Figs. 6.6 and 6.7, respectively. A scale at the top of these figures allows one to measure the flow angularity.

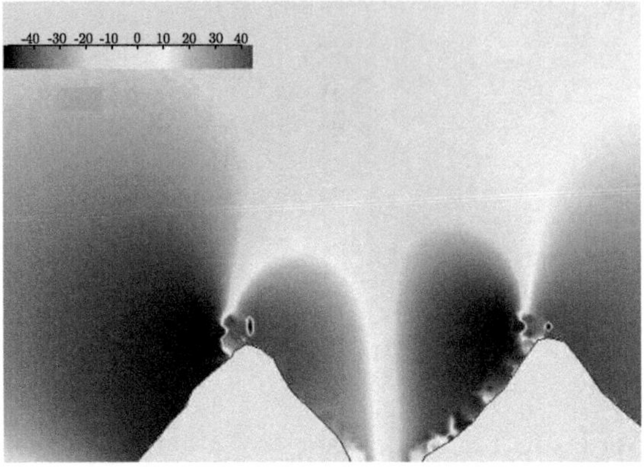

Fig. 6.6. Flow angles in a vertical plane at $\tau = 0{,}15$

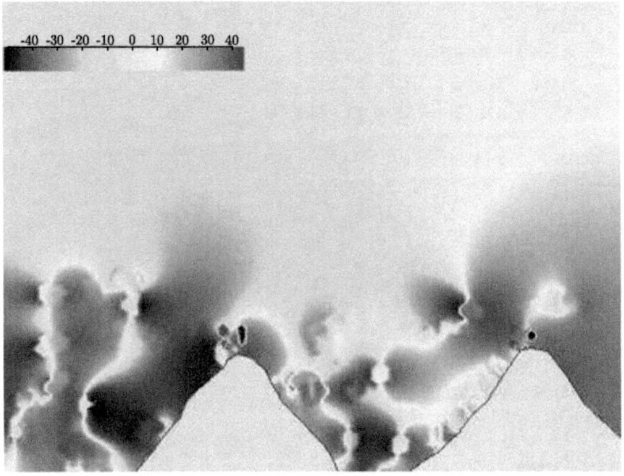

Fig. 6.7. Flow angles in a vertical plane at $\tau = 3{,}3$

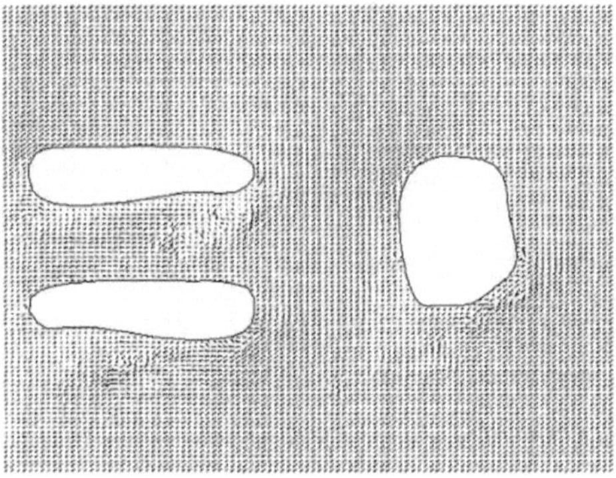

Fig. 6.8. Velocity field in a horizontal plane at $\tau = 0{,}15$

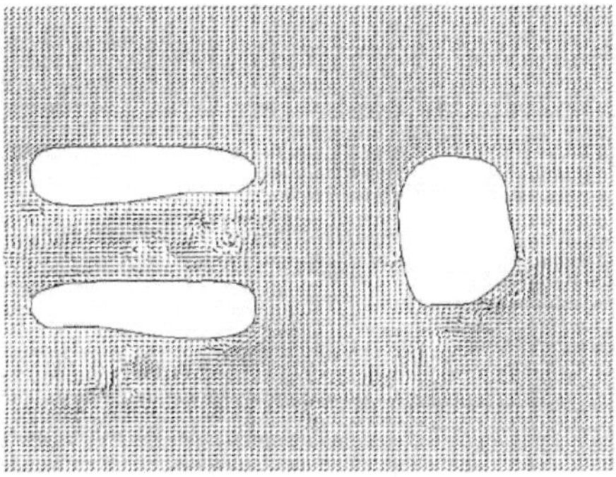

Fig. 6.9. Velocity field in a horizontal plane at $\tau = 3{,}3$

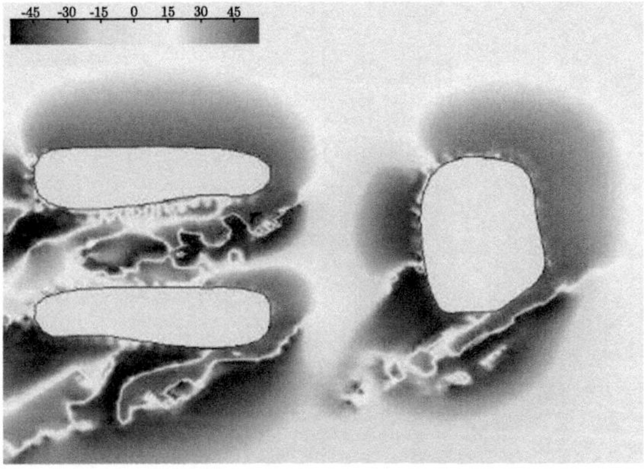

Fig. 6.10. Flow angles in a horizontal plane at $\tau = 0{,}15$

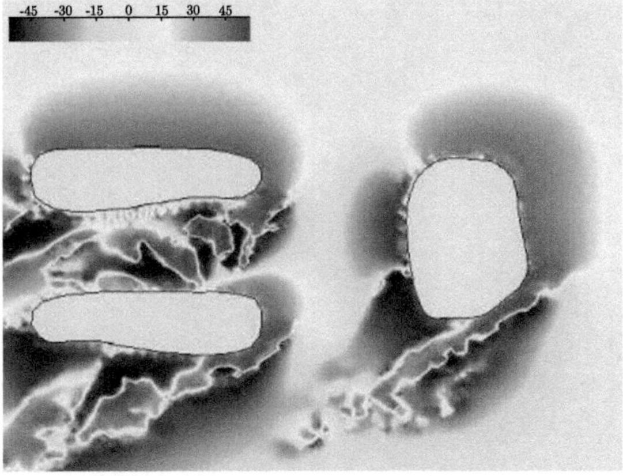

Fig. 6.11. Flow angles in a horizontal plane at $\tau = 3{,}3$

The disturbance velocity fields in a horizontal plane at nondimensional time instants of $\tau = 0{,}15$ and 3,3 are shown in Figs. 6.8 and 6.9, respectively. The horizontal plane lies at a height of 1/3 h (h is the mountain's height). The flow patterns in the two planes allow one to see the process of origination and propagation at the vortices in greater detail.

In Figs. 6.10 and 6.11 flow angles are shown in a horizontal plane at $\tau = 0{,}15$ and 3,3 respectively. At the top of these figures there is also a scale to measure flow angles.

Thus, the discrete vortex method enables modeling flows over terrains. Such information is important from a standpoint of flight safety at low heights. The next chapter is devoted to the effect of a vortex wake on the behavior of an aircraft encountering a wake.

Chapter 7

SIMULATION OF THE FAR VORTEX WAKE OF AN AIRCRAFT AT TAKEOFF AND LANDING

7.1. Problem statement

For an aircraft flying at height the far vortex wake represents two parallel, sinking, oppositely rotating vortex tubes. The loss of circulation of each of the vortex tubes with time is caused by interpenetration (diffusion) between vorticities of opposite signs. When flying in a turbulent atmosphere the diffusion is intensified by elevated turbulence, which further decreases the circulation of each tube. Various empirical formulas are available presently for assessing the loss of circulation at low and high levels of atmospheric turbulence [11, 46, 63].

Taking into account the interaction between the vortex wake and the ground surface in an inviscid approximation leads to the known result according to which the vortex system of an aircraft (two oppositely rotating vortices near the ground surface and two image vortices located symmetrically about the surface, forming a quadrupole [53]) is unsteady and both vortices sink and moving apart in the transverse direction.

In the framework of ideal fluid theory when solving a corresponding two-dimensional problem, a hyperbolic trajectory $y^2 + z^2 = y^2 z^2\ (0{,}5b_0)^2$, is realized, where z is the distance from the vortex center to the plane of symmetry for the vortex pair, y is the vortices' height above the ground surface, b_0 is the initial distance between the vortices.

Initial wind-tunnel investigations of the wing's vortex system near the ground board have already shown [38] that the wing-tip vortices do not only sink and move apart, as it follows from theory within the inviscid approximation, but they also rise to a certain height to subsequently perform a complex motion featuring their loop-shaped tracks in cross-flow planes. In Ref. [48] it was experimentally shown that such a complex motion is caused by separation of the boundary layer formed on the ground surface by a transverse flow induced by the wing's vortex system. Secondary vortices shed into the flow thorough boundary layer separation interact with the primary vortices, resulting in the aforementioned complex motion of the primary vortices (see also [52]).

Such a motion of the primary oppositely rotating vortex tubes near the ground board is described in the RANS framework with and without lateral drift [60, 68].

A.S. Ginevsky, A.I. Zhelannikov, *Vortex Wakes of Aircrafts*, Foundations of Engineering Mechanics,
DOI 10.1007/978-3-642-01760-5_7, © Springer-Verlag Berlin Heidelberg 2009

Computations were performed using the Reynolds equations closed with a differential turbulence model [60, 68], as well as using the DVM supported by techniques of the turbulent boundary layer theory [43]. In the first case the two vortex tubes were viscous (turbulent), in the second case the problem was treated in the framework of an ideal fluid with taking into account the near-wall transverse flow accompanied by the formation and subsequent separation of the turbulent boundary layer.

Fig. 7.1 presents the parameters of the vortex pair's model to numerically compute the primary vortices' motion near the flat ground surface using the Reynolds equations (Fig. 7.1 *a*) and the DVM (Fig. 7.1 *b*).

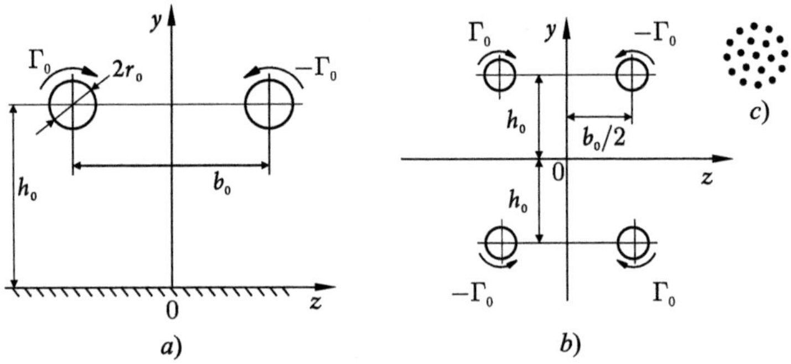

Fig. 7.1. The parameters of the vortex pair's models to numerically simulate the vortices' motion near a flat ground board using the Reynolds equations (*a*) and the DVM (*б*); *b*; *c* — the model of the vortex tube.

Fig. 7.2 demonstrates the primary vortices' motion, obtained by numerically solving the Reynolds equation with a turbulence model using a technique from Ref. [68]. In Fig. 7.2 *a* symbol *l* corresponds to the laminar regime (the solution is obtained using the Navier–Stokes equations), symbol *t* — turbulent regime (the Reynolds equations were solved with a technique of Ref. [60]). The dashed line denotes the exact solution for a vortex filament, circles denote the solution for an inviscid vortex tube (both solutions correspond to the ideal fluid case), the solid line means a turbulent regime (Fig. 7.2 *b*).

Fig. 7.3 demonstrates the tracks of the primary vortices in a cross-flow control plane as well as the location of the secondary vortices at the final stage of their lifetime with account for the attenuation of both the primary and secondary vortices. The computations were performed with a combination of the DVM and methods of the turbulent boundary layer theory [43]. Both techniques provide satisfactory simulation of the complex motion of the primary vortices near the ground.

For high Reynolds numbers characteristic to the problem under consideration numerical simulation of an aircraft's far vortex wake near the ground may be performed with the DVM combined with methods of the turbulent boundary layer theory. This in part stems from the fact that the

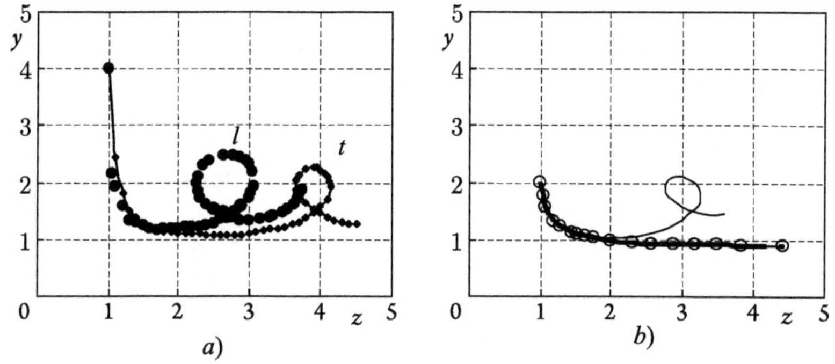

Fig. 7.2. The tracks of the primary vortices in the cross-flow control plane (l — laminar flow, t — turbulent flow). Computational methods from Refs. [60, 68]. Vertical coordinate y is nondimensionalized by h_0, horizontal coordinate z by $b_0/2$ (see Fig. 7.1).

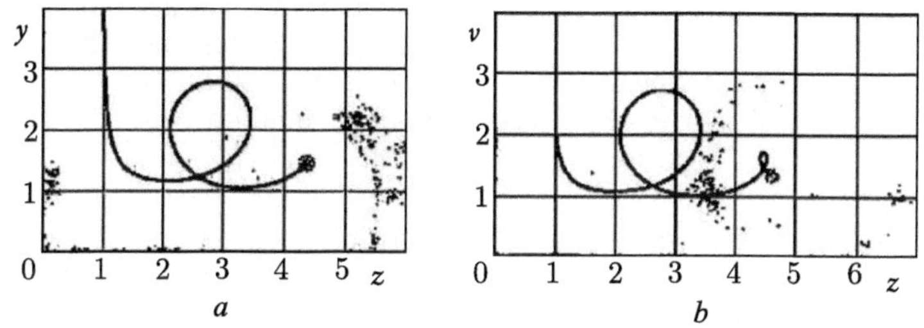

Fig. 7.3. The tracks of the primary vortices in the control plane computed with the DVM [43]. The initial locations of the primary vortices (a, b) correspond to the data of Fig. 7.2.

flow here contains the regions of concentrated vorticity (vortex tubes and boundary layer) and potential flow.

Using this basis, a simple approximate method was developed for computing the far vortex wake of an aircraft near the ground during takeoff and landing operations. The method comprises the prediction of the near vortex wake of the aircraft with its high-lift devices deployed, computation of the roll-up of the near wake into two vortex tubes and the transverse flow induced by these vortex tubes, prediction of the separation of the turbulent boundary layer combined with formation of the longitudinal (secondary) vortices and the interaction between the primary and secondary vortices. For such a problem statement there is no need to specify the diameter, coordinates of the centers and circulation of the vortex tubes — they are determined in the process of solution for the known geometry of the aircraft with the high-lift devices deployed. Taken into account in this case are the effects of the flight height, crosswind velocity and atmospheric turbulence [14–17] [1] as well as an additional diffusion source caused by viscous effects accompanied by the formation of the secondary vortices due to turbulent boundary layer separation.

The presented computation of the parameters of the aircraft far vortex wake is based on an approximation according to which the engine exhaust jets attenuate rapidly compared to the long-lived vortex wakes; because of this, the effect of the engine jets can be ignored in computing vortex wakes [40, 49]. The recent computations of the aircraft jet-vortex wakes for aircraft with different numbers and arrangements of engines have shown that the engine plumes practically do not affect the development of the far vortex wakes of aircraft [41, 51].

7.2. Simulation of an aircraft's near vortex wake. Linear theory

Consider a steady-state attached subsonic flow of an ideal gas past an aircraft. Disturbances introduced into the flow by the aircraft are taken to be small compared to the flight speed V_∞. It is assumed that the flow outside the aircraft and its vortex wake is potential. The perturbation potential φ must satisfy the Prandtl–Glauert equation

$$(1 - \mathrm{M}^2)\partial^2\varphi/\partial \mathrm{x}^2 + \partial^2\varphi/\partial \mathrm{y}^2 + \partial^2\varphi/\partial \mathrm{z}^2 = 0. \tag{7.1}$$

Here, $\mathrm{M} = \mathrm{V}_\infty/\mathrm{a}_\infty$ — Mach number, a_∞ — speed of sound in the undisturbed medium. The pressure within linear accuracy satisfies the Cauchy–Lagrange equation

$$p - p_\infty = -p_\infty V_\infty \partial\varphi/\partial x.$$

Equation (7.1) is solved for the following boundary conditions:

a) impermeability of the aircraft surfaces;

b) no zero pressure jump across the vortex sheet;

c) velocity finiteness (Chaplygin–Zhukovsky condition) at the trailing edges of lifting surfaces;

d) disturbance attenuation at infinity.

The effect of ground proximity during takeoff/landing operations was taken into account using the mirror image of the vortex wake model.

The problem under consideration is solved with the DVM [31] which reduces it to the following. The flow about an aircraft is modeled by gasdynamic singularities located on flat basic elements [8]. On lifting surfaces the basic elements are arranged to approximate their mean-surface models. Bodies of revolution (fuselage, engine nacelles) are represented by two mutually perpendicular body projections onto the vertical and

---

[1] See also *Belotserkovsky AlS, Ginevsky AS* (1999) Interaction of the far vortex wake of an aircraft with the ground surface at takeoff and landing (in Russian). TsAGI, Preprint № 123, 20 pp

horizontal planes. Such modeling allows one to take into account the lifting properties of elongated bodies and facilitate representation of the element's geometry (see Chapter 3). Vortex sheets leave the trailing edges of the wing, horizontal tail, horizontal and vertical surfaces of the nacelles during pitching motion and the vertical surfaces of the fuselage, vertical tail in lateral motion, i.e., in the presence of a crosswind.

For numerically solving the problems under study, the continuous vortex layer is represented with discrete vortices. Each basic element is divided into panels, where the vortices and control points are positioned. The boundary conditions are met at control points lying between vortices on the basic elements in the streamwise and lateral directions.

Thus, solving the problem reduces to solving a system of algebraic equations for unknown intensities of the bound vortices (loads) on the aircraft's surface. Using the known loads, the aerodynamic derivatives of the force coefficients $c_y^{\varepsilon_i}, c_z^{\varepsilon_i}$ and dimensionless vortex circulations Γ^{ε_i}. are determined. Here, ε_i — nondimensional kinematic parameters, α (angle of attack) and ω_z (nondimensional rotational speed about the $0z$ axis) for longitudinal motion; β (slip angle), ω_x (nondimensional rotational speed about the $0x$ axis), ω_y (nondimensional rotational speed about the $0y$ axis) for lateral motion; δ — deflection angle of a high-lift device [31]. The nondimensional circulations of free vortices were determined by summing the strengths of the free vortices (with consideration for their signs) and bound vortices leaving the wing.

When computing vortex wakes of different aircraft, the number of free vortices was in the range of 60 to 90. For determining the intensities at the vortices for a specified flight regime, the lift coefficient was found from the equality between weight and lift:

$$c_y = \frac{2\,m\,g}{\rho S V_\infty^2}. \tag{7.2}$$

Since at large deflections of the flaps the validity of computations within the framework of linear theory can be doubtful, a comparison was performed between computed data and flight test results [33] for a Tu-204 aircraft out of ground effect ($H = 6{,}000$ m, $\mathrm{M} = 0{,}35$). Fig. 7.4 depicts the representation of this aircraft with flat plate elements. It has the following geometrical data: wing span $l = 40{,}88$ m, MAC $b = 4{,}11$ m, gross wing area $S = 168{,}63\,\mathrm{m}^2$, wing sweep angle $\chi = 28^\circ$ and aspect ratio $\lambda = 10$. The wing has full-span slats and twin-slotted flaps with their two segments separated by the engine nacelles. At takeoff and landing, the slats were deflected through 27°, the flaps to 18° (takeoff) and 37° (landing) with a 12% chord extension.

Computations of the aircraft's longitudinal and lateral-directional aerodynamic characteristics and vortex circulation in the near vortex wake were carried out at $\mathrm{M}_\infty = 0{,}3$ and $H = 6{,}000$ m for three configurations: cruise, takeoff and landing. Because the high-lift devices' deflections were

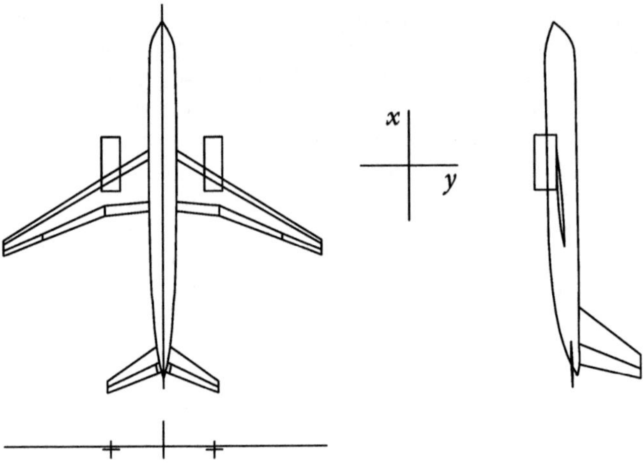

Fig. 7.4. . Plate-element representation of the Tu-204 aircraft

sufficiently large, for computing the lift coefficient the following relation was used:

$$c_y = c_y^{\alpha} \sin\alpha \cos\alpha + c_y^{\delta_{sl}} \sin\delta_{sl} + c^{\delta_{fl}} \sin\delta_{fl} \cos\delta_{fl}, \tag{7.3}$$

where δ_{sl} and δ_{fl} are deflection angles of the slats and flaps, respectively.

Compared in Fig. 7.5 are computed (dashed curves) and experimental data obtained in wind-tunnel (solid curves) and flight (points) experiments [23] for cruise (*1*), takeoff (*2*) and landing (*3*) configurations. Satisfactory agreement between theory and experiment can be seen for angles of attach $\alpha = 0 - 12°$ for the basic configuration and $\alpha = 0 - 20°$ with the high-lift devices deployed, which is probably associated with a delay in the onset of flow separation.

The flight experiment was performed at $H = 6{,}000$ m and $\mathrm{M}_\infty = 0{,}28 - 0{,}45$; in the wind-tunnel experiment the folowing test conditions were provided: $\mathrm{Re} = (1-4)\cdot 10^6$, $\mathrm{M}_\infty = 0{,}15 - 0{,}4$ (TsAGI T-106 wind tunnel) and $\mathrm{Re} = (5-6)\cdot 10^6$, $\mathrm{M} = 0{,}15$ (TsAGI T-101 wind tunnel). The wind-tunnel data presented in Fig. 7.5 are scaled down to the flight experimental conditions [33] with respect to Mach and Reynolds numbers. In computations, the deployed stats and flaps were represented with flat plates with the corresponding deflection angles and chords. The extension of the slats and flaps and the corresponding increase in the wing planform area were taken into consideration.

The comparison of the computed and wind-tunnel and flight experimental data gives ground to believe that the computed circulations are close to real data. Computational results for a B-737 aircraft [29] can serve as indirect confirmation of such a belief. This aircraft has wing span $l = 28.9$ m, wing area $S = 105\,\mathrm{m}^2$, flight speed $V_\infty = 73$ m/s, mass $m = 46{,}000$ kg, extension triple-slotted flaps. As shown in Ref. [29], the

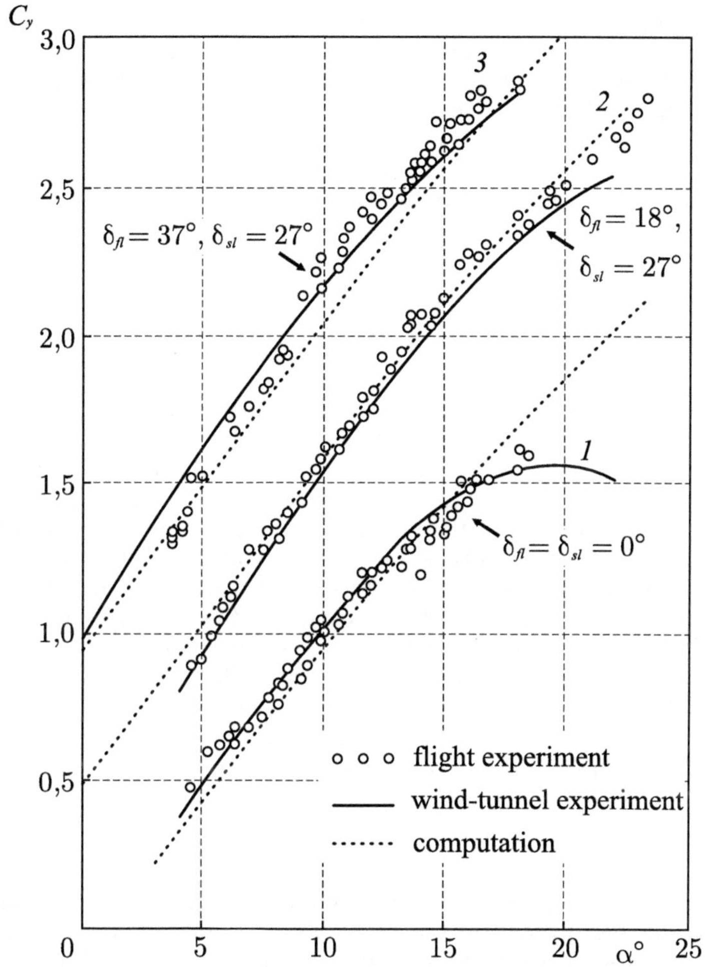

Fig. 7.5. Comparison of computed and experimental relationships $C_y(\alpha)$ for a Tu-204 aircraft for its cruise (*1*), takeoff (*2*) and landing (*3*) configurations

assessments based on experimental data give the value of circulation at landing ($H = 60\,\mathrm{m}$) equal to $\Gamma_0 = 225\,\mathrm{m}^2/\mathrm{s}$, computed data for various configuration of the high-lift devices are within the range $\Gamma_0 = 195 - 267\,\mathrm{m}^2/\mathrm{s}$. This supports the correctness of the above computed data on the circulations in the near vortex wake.

DVM computation of flow about a schematized representation of an aircraft of a specified configuration (wing, horizontal and vertical tails) in a linear approximation [7, 30] is carried out with the flaps deployed for takeoff or landing. The spanwise circulation distribution is found with allowance for the contribution of the vortex systems of the horizontal tail and the vertical tail (in the presence of a crosswind). Special studies have shown that circulation values obtained by such a way weakly depend on flight height; a significant ground effect manifests itself only at heights below the semi-span of the aircraft wing.

To determine the location of the far vortex wake of an aircraft the DVM was used [29, 30]. Computations were carried out with the known formulas derived from the Biot–Savart law:

$$u_{yi} = \sum_{j \neq i} \Gamma_j \frac{z_i - z_j}{2\pi r_{ij}^2},$$
$$u_{zi} = -\sum_{j \neq i} \Gamma_j \frac{y_i - y_j}{2\pi r_{ij}^2}, \qquad (7.4)$$

where u_{yi} and u_{zi} are the velocity components for the ith vortex, y_i, z_i, y_j, z_j are the coordinates of the ith and j^{th} vortices, respectively, $r_{ij}^2 = (y_i - y_j)^2 + (z_i - z_j)^2$, Γ_j is the circulation of the jth vortex.

Consider a model of the vortex tube in the form of a circle with uniformly distributed vorticity within it. This is provided by distribution of a large number of vortices of equal circulation. Let us compute the disturbance velocity u_{yz} induced by a given vortex tube at the horizontal straight line passing through the center of the vortex tube. According to relations (7.4), one can obtain the vertical (u_{yi}) and horizontal (u_{zi}) velocity components at points located on the line. In this case the summation is performed over all vortices, correspondingly y_i, z_i are the coordinates of points of the line, r_{ij}^2 is the square of the distance from the ith point to the jth vortex. As a result, we get the velocity u_i at the ith point of the line:

$$u_i = \sqrt{u_{yi}^2 + u_{zi}^2}\,. \qquad (7.5)$$

Expression (7.5) determines the magnitude of the induced velocity. Although in the given case the velocity's horizontal component is nearly zero, the use of expression (7.5) is justified since in computing real flights this condition is not always met. The computed graph is shown in Fig. 7.6. It should be noted that the maximum of the velocity magnitude is observed on the boundary of the vorticity region, whereas its minimum lies at its center, i.e., from this velocity graph it is possible to determine the location and size of the vortex tube.

Consider now the vortex panel representation of the near wake of the schematized aircraft model in a linear approximation. Let us make use a data base computed in advance, which contains the values of dimensionless derivatives of circulations of vortices and their coordinates. For different aircraft types the total number of vortices in the data base can be from 60 to 100. Vortex circulations are computed in accordance with flight conditions. Considered in the given case is the near wake of a B-727 aircraft flying at $H = 150\,\mathrm{m}$ (to exclude the ground effect) and

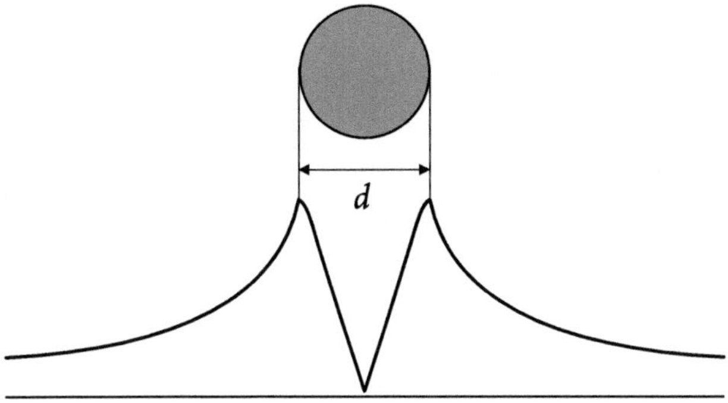

Fig. 7.6. Velocity magnitude distribution within the vortex tube and outside it

$V = 72\,\text{m/s}$, with an angle of attack of 8.1° and the flaps deflected through 25°. The vortex scheme of the near wake is depicted at the top of Fig. 7.7 *a*. From here on the roll-up of the wake into two vortex tubes occurs – one on the right side is denoted by subscript 1, that on the other side by subscript 2. The vortices formed on the vertical tail are designated by subscript 0.

For determining the induced velocity it is necessary to know the location of the two centers of the vortex system, whose coordinates were computed as the ratio between the sum of the products of the circulations of each vortex of the vortex tube by the corresponding coordinates and the sum of these circulations. Vortices generated on the vertical tail under the action of a crosswind can be referred equally to both vortex tubes and, because of this, when computing the coordinates of each of the centers, the correction was added equal to one-half the contribution of these vortices. Thus, we get the following formulas for the right vortex tube:

$$y_{1c} = \frac{\sum_i \Gamma_{li} y_{li}}{\sum_i \Gamma_{li}} + 0{,}5 \frac{\sum_i \Gamma_{0i} y_{0i}}{\sum_i \Gamma_{0i}}, \qquad (7.6)$$

$$z_{1c} = \frac{\sum_i \Gamma_{li} z_{li}}{\sum_i \Gamma_{li}} + 0{,}5 \frac{\sum_i \Gamma_{0i} z_{0i}}{\sum_i \Gamma_{0i}},$$

and similar formulas for the left tube. To meet the tangency condition for the ground surface, the scheme with image vortices is used in the computation. The vortices, the real basic and image ones, are located symmetrically to the ground surface and have opposite signs.

The induced velocity whose plot is given in Fig. 7.7 *a* is obtained from relations (7.4) and (7.5) in a similar manner to that done in the previous case. It can be seen that the plot has several maxima and minima:

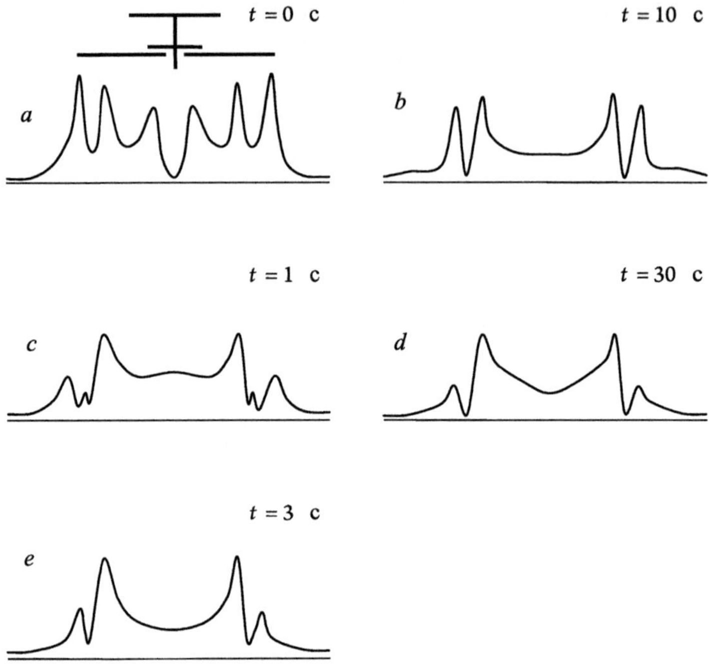

Fig. 7.7. Evolution of the near vortex wake behind a B-727 aircraft: the profiles of the velocity magnitude at the initial instant of time *(a)*, at $t = 1$ c *(b)*, $t = 3$ c*(c)*, $t = 10$ c *(d)* and $t = 30$ c *(e)*

the reason is that the vortex wake tails not only the wing but also the deflected flaps, and that the vortex sheet at the beginning of computation ($t = 0$) has not still rolled up into the vortex tubes.

Let us consider the further evolution of the wake. The general pattern of flow has changed after one second (Fig. 7.7 *b*). In the velocity plot one can distinctly observe four maxima (two at the left and two at the right) and two small intermediate maxima which is caused by deflection of the flaps and the fact that the wake has not still completely rolled up. It should be noted that all vortices have different circulations, and in this case eight vortices located at the right and symmetrically at the left have the largest circulation and to a large extent determine the behavior of the vortex system as a whole.

After 3 s subsequent to the beginning of the roll-up, the intermediate maxima almost disappear (Fig. 7.7 *c*) and by $t = 10$ s the vortex wake can be considered fully rolled up (Fig. 7.7 *d*). From this point on the computation reveals no significant qualitative change in the distribution of the velocity induced by the aircraft vortex wake. Fig. 7.7 *e* demonstrates the vortex wake and velocity distribution at $t = 30$ s.

7.3. An approximate computation of an aircraft's far vortex wake

The vortex sheet leaving an aircraft in flight is unstable: it rolls up into two oppositely signed vortex tubes. This problem cannot be solved in the framework of linear theory. However, an approximate technique [8, 73] may be used for this purpose: the vortex sheet leaving the aircraft is reckoned two-dimensional, i.e., invariable along the longitudinal coordinate $-\infty < x < \infty$ (see Chapter 3). In this case its roll-up into two vortex tubes can be computed as in solving a two-dimensional unsteady problem. As a result, we get the time evolution of the vortex tubes in the far vortex wake. Besides, we obtain the parameters of this tubes: their shape and the distribution of the velocities u_y and u_z inside them.

Thus, the steady process of rolling up the three-dimensional vortex sheet is reckoned equivalent with the unsteady roll-up process of the two-dimensional vortex sheet in time t, $t = x/V_\infty$ [73]. To determine the velocity induced by the vortex tube at distances large compared to its diameter, it is necessary to determine the locations of the two centers of the vortex system (1.1), (2.1) whose coordinates are computed as the ratio between the sum of products of the circulation of each of the vortex filament forming the vortex tube by the corresponding coordinates and the sum of these circulations:

$$y_{1c} = \frac{\sum\limits_i \Gamma_{li}\, y_{li}}{\sum\limits_i \Gamma_{li}\, z_{1y}}, \qquad z_{1c} = \frac{\sum\limits_i \Gamma_{li}\, z_{li}}{\sum\limits_i \Gamma_{li}}. \tag{7.7}$$

The computation for the second vortex tube is performed similarly.

The numerous experimental and theoretical studies testify that the circulations of the vortex tubes gradually decrease with time. This so-called loss of circulation is caused, as was indicated in Introduction, by diffusion of vorticity of the two oppositely signed vortex tubes. To determine the time dependence of the circulation, one can use the empirical formula [46, 48, 63]

$$\frac{\Gamma_1(t)}{\Gamma_o} = \left[1 - \exp\left(-\frac{r_i^2}{4\nu_{i1} t}\right)\right] \exp\left(-c\frac{qt}{b_1}\right), \tag{7.8}$$

where Γ_i and Γ_0 are respectively the current and initial values of the circulation of the primary vortices, t is time, q is the root mean square of the velocity fluctuation in the atmosphere, $b_1(t)$ is the distance between the centers of primary vortices 1 and 2 along the z axis, $r_i = \left[(z - z_i)^2 + (y - y_i)^2\right]^{1/2}$ is the magnitude of the position vector of the point with the coordinates z, y relative to the vortex with the coordinates z_i, y_i, ν_{i1} is the eddy viscosity coefficient, an empirical multiplier $C = 0.41$ [63].

The first multiplier in expression (7.8) takes into account a decrease in circulation at low atmospheric turbulence, the second multiplier allows for the loss of circulation caused by elevated atmospheric turbulence. The mean value of ν_i for the vortex tube is about $0.25\,\mathrm{m}^2/\mathrm{s}$ according to computations and processing of field experiments' data [9]. Subscript i denotes the number of the corresponding vortex filament in the vortex tube.

The first multiplier in formula (7.8) can be also used for the case of aircraft flying near the ground, however, with approaching the ground the primary vortex tubes move apart, i.e. the distance between them, b_1, increases along the z axis, which must be taken in account in computation.

7.4. Generation of crossflow by vortex tubes. Turbulent boundary layer computation

The vortex tubes of aircraft flying at low heights, $H < 100\,\mathrm{m}$, induce in the aerodrome's zone a crossflow (initially converging, then diverging), which influences the development of the boundary layer and its separation both in the absence and presence of a weak crosswind. Taking into account the roughness of the runway and a long distance reaching 10 – 20 m on which this process occurs, it is reasonable to suppose that this layer is fully turbulent.

With no crosswind, the distribution of the velocity $u_z(z)$ induced by the primary vortices near the aerodrome's surface is symmetric. With a crosswind, the flow pattern becomes asymmetric and the computation of the boundary layer is carried out on each side of the critical point where $u_z = 0$. To a first approximation, one can consider that $u_{z\delta}(z) = u_z(z)$. The wind speed can be taken constant along the vertical coordinate and the effect of the wind is limited to the drift of the vortex system in the direction of the wind.

To determine the circulation of the secondary vortices it is necessary to compute the parameters of the boundary layer at the separation station, z_{sep}. For this purpose, one of the integral methods can be used, for example, the method from Ref. [39], based on a polynomial representation of the tangential stresses' profile and the use of the Prandtl formula $\tau = \rho l_o^2 (\partial u/\partial y)^2$, где l_o is the mixing length. The computational procedure consists of the numerical integration of the momentum equation and two interpolating expressions of closure for the surface friction coefficient $c_f = 2\tau_w/\rho u_{z\delta}^2$, and the formparameter $H = \delta_1/\delta_2$ of the boundary layer (here δ_1 and δ_2 are the displacement and momentum thicknesses, respectively, u is the velocity in the boundary layer, τ_w is the tangential stress of the surface friction). The relation $u_z(z)$ involved in the indicated system of equations can be computed from the known parameters of the vortex tubes in the aircraft far vortex wake for a specified flight height.

Thus, the sought-for formulas are:

$$\frac{dR_2}{d\overline{z}} = \frac{1}{2}\mathrm{Re}\overline{u}_{z\delta}c_f - \frac{\overline{u'}_{z\delta}}{\overline{u}_{z\delta}}(H+1)\,R_2, \tag{7.9}$$

$$c_f = c_{fo}[1 + \lambda_1 f + \lambda_2(e^{\lambda_3 f} - 1)], \tag{7.10}$$

$$H = H_o\,(1 - \lambda_4 f) - 0{,}019 f e^f \xi. \tag{7.11}$$

Here,

$$\lambda_1 = 0{,}2814 - 0{,}036\xi + 3{,}6\xi^{-4,5},$$
$$\lambda_2 = 0{,}1185\xi - 0{,}262,$$
$$\lambda_3 = 0{,}585 - 0{,}125\xi + 20{,}4\xi^{-1,75},$$
$$\lambda_4 = 0{,}28 - 0{,}034\xi + (0{,}1\xi)^9,$$
$$c_{fo} = 2ce^{-0,391\xi},$$
$$H_o = 1{,}251 - 0{,}0131\xi + 5{,}35\xi^{-2,85},$$
$$c = 0{,}001[6{,}55 - 0{,}0685(\xi - 4{,}4) + 0{,}256(\xi - 4{,}4)^2];$$

$$\mathrm{Re} = \frac{V_\infty l}{\nu}; \qquad R_2 = \frac{u_{z\delta}\delta_2}{\nu}; \qquad c_f = \frac{2\tau_w}{\rho u_{z\delta}^2};$$

$$\xi = \lg R_2; \qquad f = \frac{e^{2,694\xi}\overline{u'}_{z\delta}}{c\mathrm{Re}\overline{u}_{z\delta}^2}; \qquad \overline{u'}_{z\,\delta} = \frac{d\overline{u}_{z\delta}}{d\overline{z}},$$

l is the wing's semi-span; V_∞ is the flight speed; $\overline{u}_{z\delta} = u_{z\delta}/V_\infty$, $\overline{z} = z/l$, $\delta_1 = \int_0^\delta (1 - u/u_{z\delta})dy$, $\delta_2 = \int_0^\delta (u/u_{z\delta})\,(1 - u/u_{z\delta})\,dy$ are the displacement and momentum thicknesses, respectively; u is the velocity in the boundary layer.

To determine the boundary layer's thicknesses δ, δ_1 and δ_2 at the separation station ($c_f = 0$), the following approximate formulas can be used:

$$\frac{u_z}{u_{z\,\delta}} = \left(\frac{y}{\delta}\right)^{\frac{H-1}{2}} \quad \text{and} \quad \frac{\delta_2}{\delta} = \frac{H-1}{H\,(H+1)}, \quad \frac{\delta_1}{\delta} = \frac{H-1}{H+1}. \tag{7.12}$$

The drift rate of the vorticity centroid of the separated boundary layer at a distance $y = \delta_1$ from the wall is

$$\frac{u_z(\delta_1)}{u_{z\delta}} = \left(\frac{H-1}{H+1}\right)^{\frac{H-1}{2}}.$$

Taking $H \approx 2$ at the separation station, we get

$$\frac{\delta_2}{\delta} = \frac{1}{6}, \quad \frac{\delta_1}{\delta} = \frac{1}{3}, \quad \frac{u_z(\delta_1)}{u_{z\delta}} = 0{,}577.$$

The vorticity of the separated boundary layer is determined from the known expression

$$\frac{d\Gamma}{dt} = \int_0^\delta \frac{\partial u_z}{\partial y} u_z dy = \frac{1}{2} u_{z\delta}^2,$$

from where we get

$$\Gamma_i = \frac{1}{2} \int_{t_i}^{t_i+\Delta_i} u_{z\delta}^2 dt. \quad (7.13)$$

The longitudinal velocity of the vorticity drift is equal to the local velocity in the boundary layer at the separation station, $y = \delta_1$, and eddy viscosity coefficient

$$u_{z\delta(y=\delta_1)} = 0{,}577(u_{z\delta})_{\text{sep}}, \quad \nu_i = l_o^2 \frac{\partial u}{\partial y} = k(\delta_2 u_{z\delta}), \quad k = 1{,}24 \cdot 10^{-3}. \quad (7.14)$$

Assume that an expression similar in structure to (7.7) can be derived for the separation station of the turbulent boundary layer, too, with the only difference that it will involve the value of the characteristic eddy viscosity at the point of vorticity shedding (7.14). Then instead of formula (7.7) we get

$$\Gamma(t) = \frac{1}{2} \left(u_{z\delta}^2\right)_{\text{sep}} K\Delta t, \quad K(t) = 1 - \exp\left[-\frac{y^2 + (z - z_{\text{sep}})}{4(\nu_i)_{\text{bl}} t}\right]. \quad (7.15)$$

Here subscripts *sep* and *bl* mean separation and boundary layer. In this case the loss of the secondary vortices' circulation is caused by viscous effects accompanying the formation of these vortices.

7.5. Computation of the far vortex wake behind the B-727 aircraft with account for the effect of the boundary layer on an aerodrome's surface. Comparison between computational results and flight test data

For solving the problem at hand, a procedure based on the method discussed in Section 7.4 has been introduced into the computer program. At each integration step the location of the vortex tubes' centers were determined, then the velocity $u_{z\delta}$ distribution on the ground surface was computed (along the $0z$ axis in Fig. 7.8). In this case it was assumed that $u_{z\delta} = u_z$ $(y = 0)$.

To meet the tangency condition, as noted earlier, the vortex system was supplemented with image vortices, Γ_1', Γ_2' (symmetrically to the primary vortices about the ground surface) and Γ_3', Γ_4' (symmetrically to the secondary vortices). The complete computational scheme of the vortex system is shown in Fig. 7.9. The number of the primary vortices

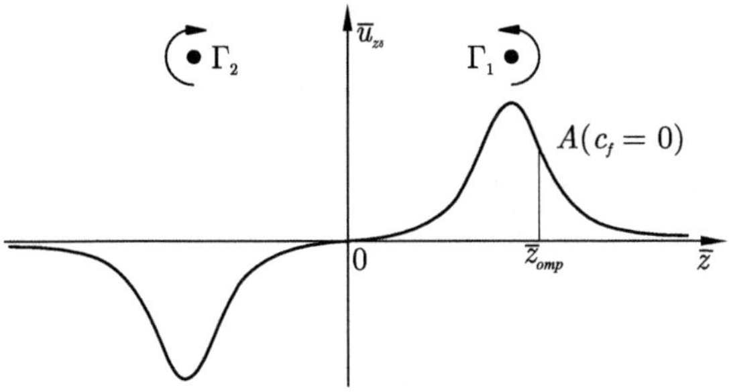

Fig. 7.8. Distribution of the velocity $u_{z\delta}(z)$

was 62, the number of the secondary vortices reached 120 by the end of computation.

In accordance with the described methodology, the computation of the far vortex wake of the B-727 aircraft flying at heights $H = 40$ m, 60 m, 80 m was computed with allowance for the effect of the boundary layer on the ground. One run of a PC takes about one minute to compute the near and far vortex wakes for 120 s of their lives at a specified flight height, crosswind velocity and degree of atmospheric turbulence. Similar computations based on the large eddy method (Navier–Stokes equations plus subgrid turbulence model) for a specified geometry and circulation of the vortex tubes require the use of a supercomputer [9]. When using for this purpose a cluster of high-speed PCs, the computational time exceeds 1000 h [71].

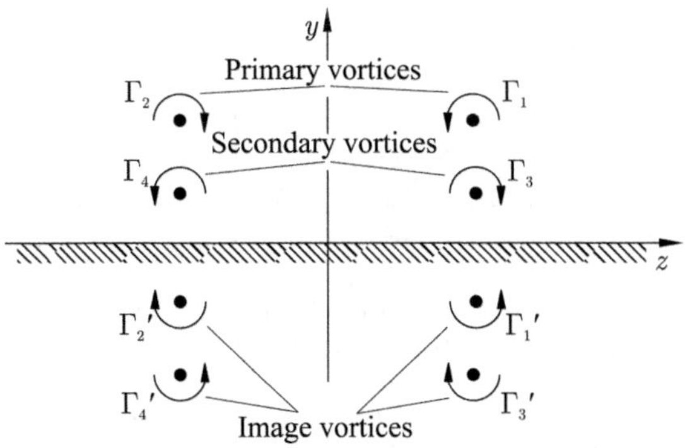

Fig. 7.9. Configurational scheme of the vortex structure: primary, secondary and image vortices

Fig. 7.10 demonstrates the results of computation of the far vortex wake behind a B-727 aircraft flying with a landing speed of 79 m/s at a height of 40 m with (*a*) and without (*b*) a weak crosswind $V_w = 1$ m/s (the flaps were set at $\delta_{\text{fl}} = 25°$, angle of attack was 5.6°).

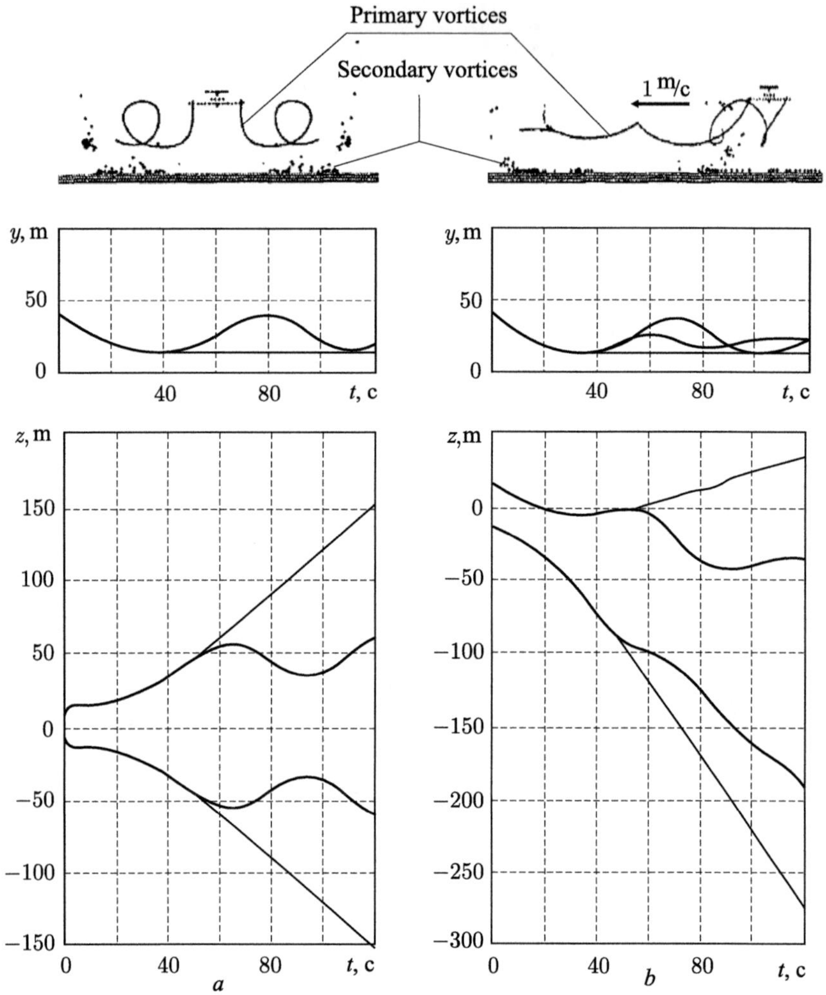

Fig. 7.10. Computed vortex wake of a B-727 at landing at $H = 40$ m with no wind (a) and a weak crosswind of 1 m/s (b) [14].

In the figure there are the time histories of the tracks (y and z coordinates) left in the control plane by its intersection with the primary vortices over the 120 s of flight time. Shown at the top are the tracks of the primary vortices in the cross-flow control plane (rear view) at different instants of time, as well as those of the secondary vortices causing the deformation of the primary vortices. The height of vortices' rebound at $t = 80$ s reaches 20 m. Thin lines of the plots $y(t)$ and $z(t)$ demonstrate the time history of these coordinates for an inviscid gas when there are no secondary vortices. Similar computations were carried out for flight

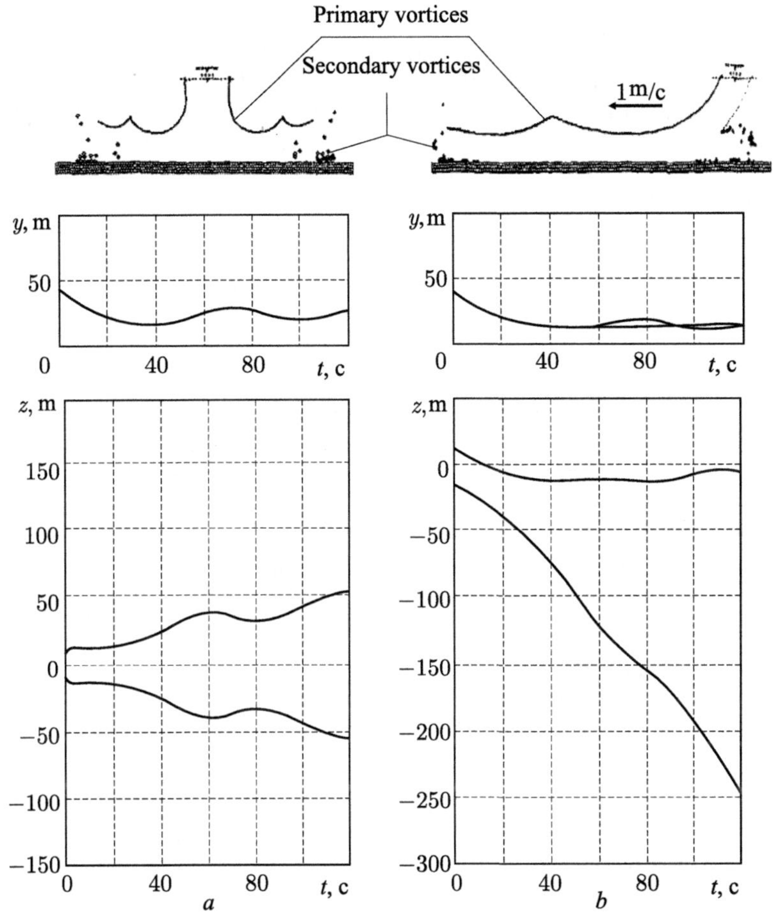

Fig. 7.11. Computed vortex wake of a B-727 at landing at $H = 40$ m with no wind (*a*) and a weak crosswind of 1 m/s (*b*) in a turbulent atmosphere with $\varepsilon = 1{,}5\%$ [15]

heights 60 m and 80 m, where the rebound height was about 30 m at $t = 95$ s and 120 s.

Similar computations are of interest for the same aircraft at a high degree of atmospheric turbulence, $\varepsilon = q/V_\infty = 1.5\,\%$. The computational results for $H = 40$ m are shown in Fig. 7.11; from these data it follows that elevated turbulence results in a significantly lower rebound and a decreased distance between the vortex tubes.

It is also of interest to compare the computed data with results of flight experiments with measurement of a vortex wake's parameters during takeoff and landing operations. Let us use the data obtained in

1995 at Memphis airport (USA) from experiments with a B-727 aircraft [35]. Unfortunately, from the material for low flight heights only one experiment had satisfactory measurements. The data from this experiment used for computation by the methodology presented in the monograph are as follows: gross weight 63,950 kg, flight height 34.8 m, flight speed 79.2 m/s, wind velocity 1.3 m/s, flap setting 25°, angle of attack 5.6°.

A comparison of computed data with the field measurement is given in Fig. 7.12. One can see satisfactory agreement between theory and experiment. Using this methodology, vortex wakes were computed for the B-727, B-747, B-777, A-300 aircraft as well as for the Russian — built Tu-204 and Il-96, when flying at heights of 20 – 80 m.

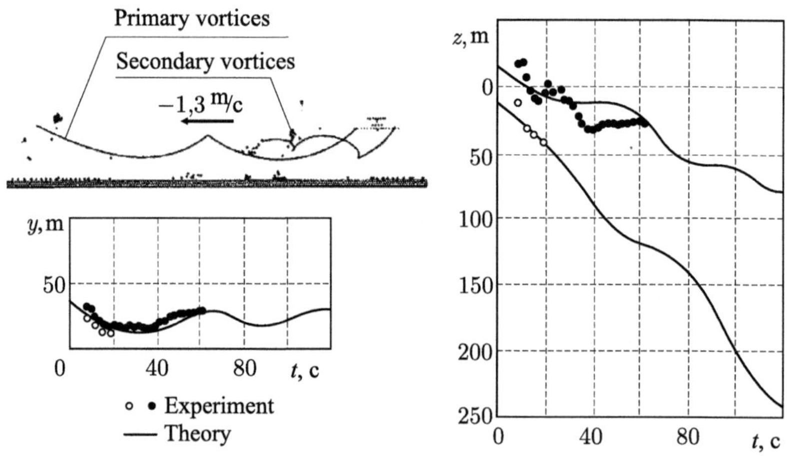

Fig. 7.12. Comparison of computed data and a flight experiment's results for a B-727 aircraft B-727 ($H = 34{,}8\,\mathrm{m}$, $V_w = -1{,}3\,\mathrm{m/s}$) [35]

7.6. Computation of the far vortex wake of Russian–built Tu-204 and Il-96 trunk-route aircraft at landing

Computation of the vortex systems of the Tu-204 and Il-96 aircraft at landing was carried out with the aim to illustrate the effect of flight height ($H = 20$ – $80\,\mathrm{m}$) and crosswind on the deformation of the vortex tubes (primary vortices).

Using the known mass of the aircraft and the computed dependence $c_y(\alpha) = c_y^\alpha \alpha$ at specified slat and flap settings, the initial values of the circulation of the vortex tube, Γ_0, the wing's angle of attack α and the landing speed V_∞ were determined. The following aircraft characteristics were used in the computations: for the Tu-204 — gross mass 100 t, angle of attack $\alpha = 8°$, circulation $\Gamma_0 = 282\,\mathrm{m}^2/\mathrm{s}$, landing speed $V_\infty = 70\,\mathrm{m/s}$;

for the Il-96 — gross mass 200 t, angle of attack $\alpha = 8\,^{\circ}$, circulation

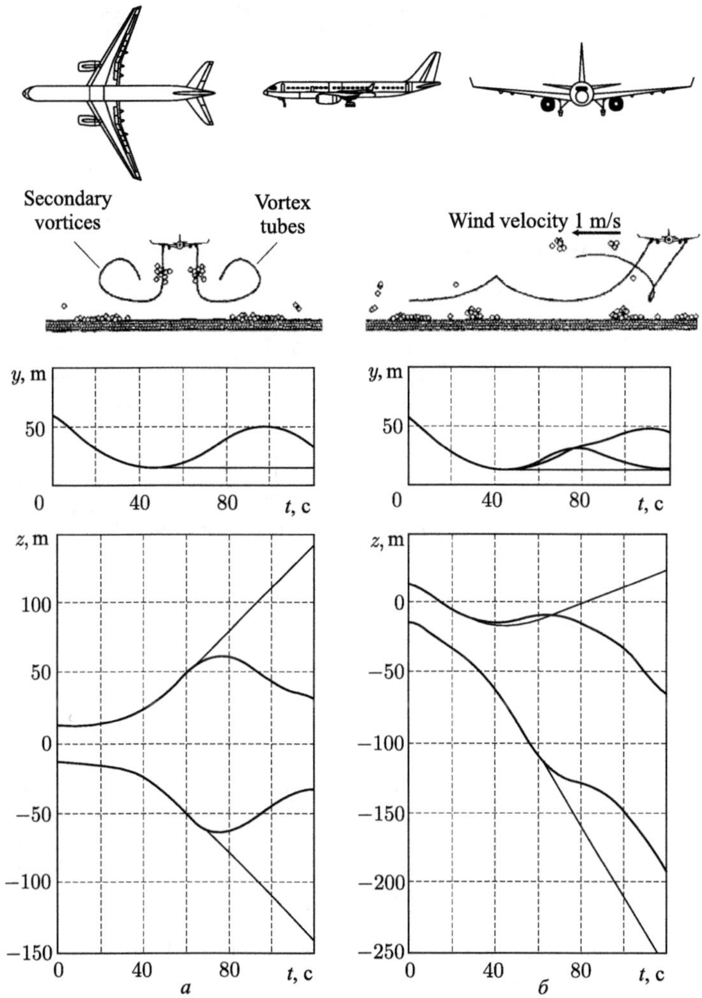

Fig. 7.13. Evolution of the vortex wake of the Tu-204 aircraft at landing ($H = 60\,\mathrm{m}$, $V_w = 0$ и $1\,\mathrm{m/s}$)

$\Gamma_0 = 422\,\mathrm{m}^2/\mathrm{s}$, landing speed $V_\infty = 85\,\mathrm{m/s}$.

Presented in Fig. 7.13 are the time histories of the vortex tubes' tracks, $y(t)$, $z(t)$, in the control plane for the Tu-204 aircraft when flying at $H = 60\,\mathrm{m}$ with no wind (a) and with a weak crosswind (b) for the first 120 s of flight. Fig. 7.14 demonstrates similar data for the Il-96 aircraft [17].

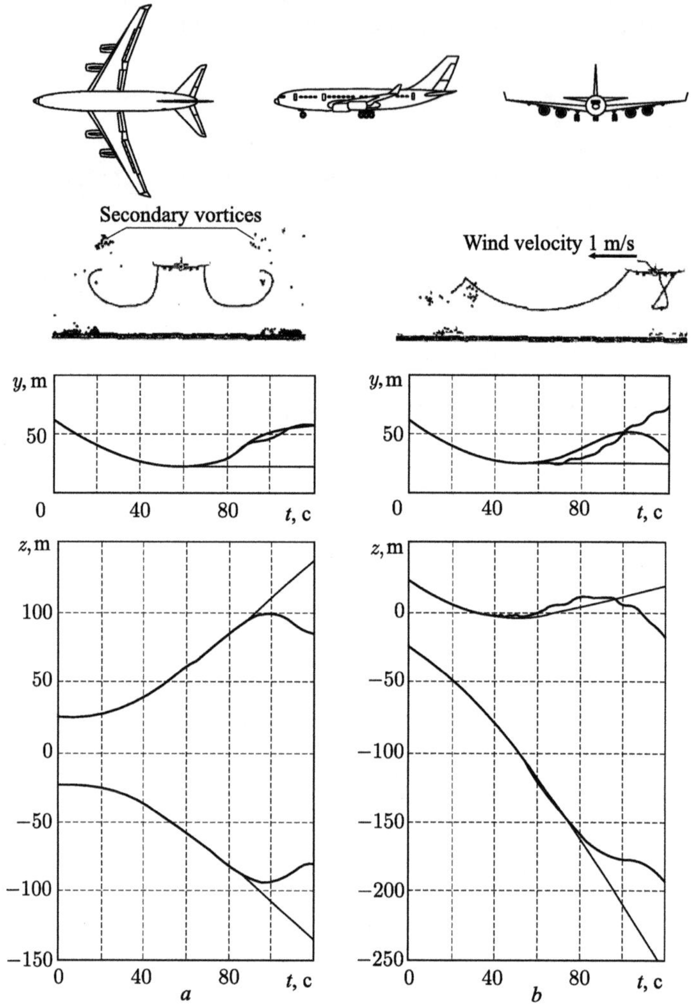

Fig. 7.14. Computation of the far vortex wake of the Il-96 aircraft at landing ($H = 60\,\text{m}$, $V_w = 0$ и 1 m/s)

The data presented provide a qualitative and, to some extent, quantitative insight into the development of the vortex wake behind the Tu-204 and Il-96 modern transport aircraft on landing.

7.7. On the visualization of an aircraft's far vortex wake near the ground

Visualization of the far vortex wake is an issue of great practical interest. It would allow pilots to see during takeoff and landing operations the picture of the primary vortex tubes, their configuration, rebound and lateral drift in the presence of a crosswind. Visualization can be realized by introduction into the near vortex wake of solid particles. Accordingly, when computing the flow with the algorithm developed earlier, solid particles are introduced into the near vortex wake and their movement is tracked in the process of rolling up the vortex sheet into the two vortex tubes of the far wake.

It is assumed that the particles do not interact with one another and do not affect the flow whose parameters are known at each instant of time, their equations of motion take into account their drag and gravitational forces. The drag coefficient of a spherical particle is expressed with a simple single-term formula $c_R = a\mathrm{Re}^{-n}$. [59]. Here, the Reynolds number $\mathrm{Re} = \rho_1 ud/m_1$, where $u = \left[(v_{1z} - v_{2z})^2 + (v_{1y} - v_{2y})^2\right]^{1/2}$ is the magnitude of the velocity vector; subscripts 1 and 2 are related to the flow and particle velocities, respectively; ρ_1 and m_1 are the air density and viscosity coefficient, d is the particle's diameter, coefficients a and $n$are known empirical function of Reynolds number. In the problem under study, the particles' density is significantly exceed the air density, i.e., $\rho_2 \gg \rho_1$.

The equations describing the motion of particles along the $0y$ and $0z$ coordinate axes are as follows:

$$\frac{dv_{2y}}{dt} = \frac{3c_R\rho_1}{4d\rho_2}(v_{1y} - v_{2y})\left[(v_{1z} - v_{2z})^2 + (v_{1y} - v_{2y})^2\right]^{1/2} - g,$$

$$\frac{dv_{2z}}{dt} = \frac{3c_R\rho_1}{4d\rho_2}(v_{1z} - v_{2z})\left[(v_{1z} - v_{2z})^2 + (v_{1y} - v_{2y})^2\right]^{1/2}.$$

where g is the gravitational acceleration.

Prior to computation, the coordinates of particles of a specified diameter d were fed into the computer program. The particles were uniformly distributed along the wing's trailing edges and their velocities V_{2z} and V_{2y} were specified. The motion of these particles was then followed in the process of the vortex sheet's rolling up into the two primary vortex tubes. At every instant of time the coordinates of the centers of mass of the solid particles' clouds were computed as well as centers of the vortex tubes, which were determined as the ratio between the sum of the product of the circulation Γ_i of each of the longitudinal vortices inside the tube by the corresponding coordinates y_i and z_i, and the sum of these circulations, $\sum \Gamma_i$. The number of vortex filaments with the circulation Γ_i behind each of the half-wing was from 60 to 90. In computing the unsteady process of the sheet's roll-up into the two vortex tubes the time step was taken to be

$\Delta t = 0.2$ s. When numerically modeling the transport of the solid particles the time step $\Delta t = 0.2$ s was decreased by a factor of 25 to 40, and with decreasing the particles' diameter this time step was decreased too.

The numerical visualization was carried out of the far vortex wake behind the Tu-204 aircraft when flying at $H = 20 - 80$ m with a weak crosswind of $V_w = 1$ m/s and under no-wind conditions ($V_w = 0$) for various particle diameters ($d = 10 - 50\,\mu$m) and densities ρ_2. One run at specified H, V_w, ρ_2 and d took $120 \div 150$ s on a PIY-1700 computer.

Fig. 7.15 demonstrates the time histories of the wake's track in the control plane in the form the y and z coordinates of the primary vortex centers and the centers of mass of the solid particles' cloud at $H = 20 - 80$ m, $V_w = 0$ and 1 m/s, $d = 10 - 50\,\mu$m. Shown at the top of the figure are the tracks of the primary vortices, secondary vortices and the centers of mass of the solid particles' clouds (rear view).

Consider now the effect of the diameter d and density ρ_2 of the particles on the quality of visualization of the far vortex wake of an aircraft under the following conditions: $H = 80$ m, $V_w = 0$ and 1 m/s, $d = 10$, 30 and 50 μm at $\rho_2 = 1000\,\text{kg/m}^3$ (see Fig. 7.15). At $d > 25\,\mu$m the center of mass of the particle cloud significantly deviates from the centerlines of the vortex tubes (*b*), which is caused by the effect of gravity force. Similar computations for $d = 10\,\mu$m and $\rho_2 = 1000 - 4500\,\text{kg/m}^3$ show that in this case the effect of gravity is weaker.

Thus, for visualization of the vortex wake of an aircraft it is advantageous to use smaller solid particles, $d \leqslant 10 \div 15\,\mu$m. This conclusion is consistent with an earlier finding of Ref. [89], according to which for visualization of a Lamb vortex in the gravitational filed particles of 10 – 20 μm in diameter may be used. But the question of the physico-chemical properties of the particles providing the wake's visualization still remains open [47].

7.8. Conclusions and prospects

Presented in this section is an effective approximate method for modeling the far vortex wake of heavy aircraft at takeoff and landing, which is based on a combination of the discrete vortex method and the integral method for computing turbulent boundary layers. An important advantage of the proposed method is the allowance for peculiarities of an aircraft, in particular, for its high-lift devices, in computing the near vortex wake and the roll-up of the wake into the two vortex tubes of the far vortex wake. Computations have been carried out for a number of Russian — built and western trunk-route aircraft at different conditions (flight height, wind velocity, atmospheric turbulence). Computed data for the lift coefficient, vortex tubes' initial circulation and their motion are in satisfactory agreement with the known data of wind-tunnel and flight experiments.

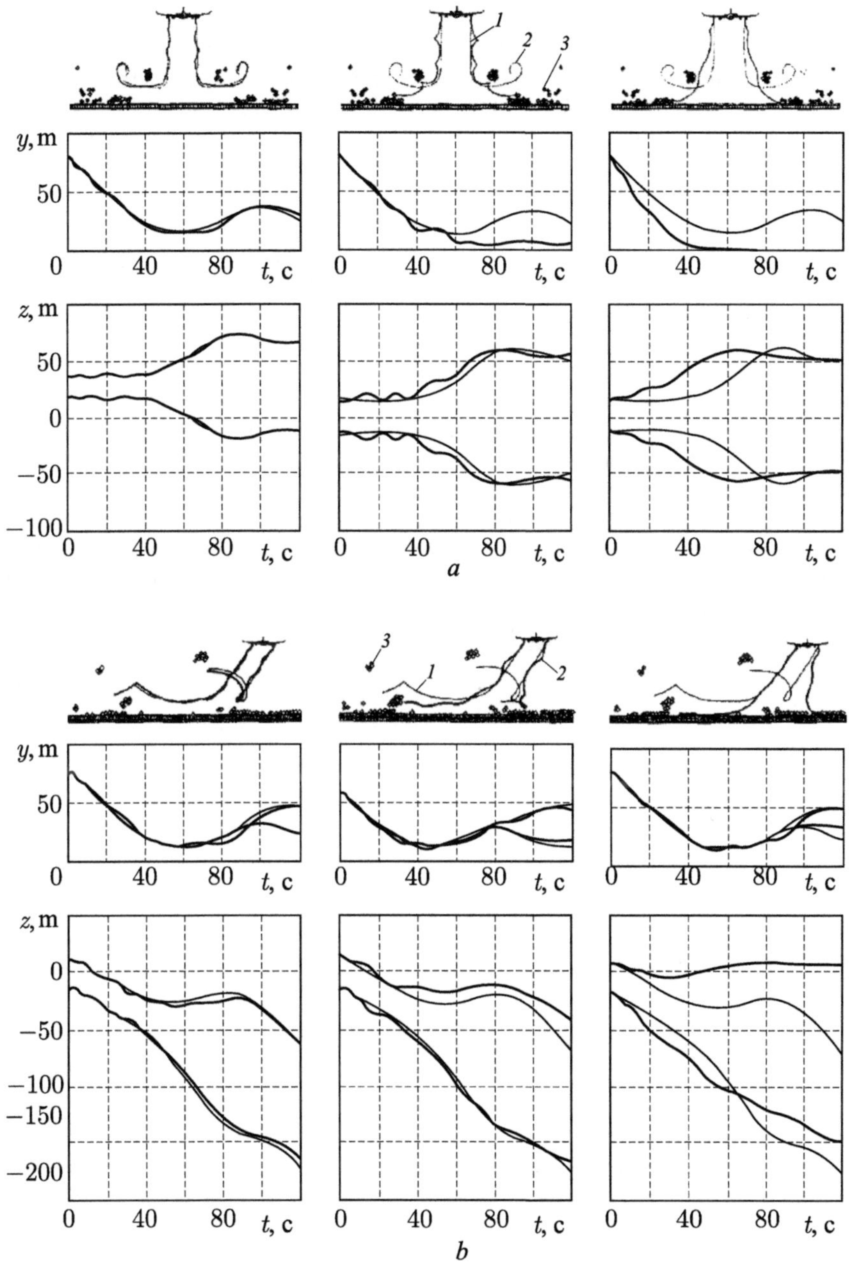

Fig. 7.15. The motion of the vortex tubes and the centers of mass of the solid particles' clouds in the wake behind the Tu-204 aircraft ($H = 80\,\mathrm{m}$, $V\mathrm{w} = 0$ and 1 m/s, $\rho_2 = 1000\,\mathrm{kg/m^3}$, $d = 10{,}3$ и $50\,\mu\mathrm{m}$): *1* — vortex tubes, *2* — centers of mass of solid particles' clouds, *3* — secondary vortices [16, 18].

In particular, it is possible in the framework of the DVM to study the interaction of an aircraft's vortex wake with the ground surface at takeoff and landing in the case when the wind velocity varies with height is accordance with the Monin–Obukhov theory or experimental data for the near-ground atmospheric layer at stable, neutral and unstable stratification [34, 58]. For this purpose, modeling a two-dimensional flow is required, invariable along the Ox axis and directed along the 0z axis.

This flow at $y > 0$ can be represented by a system of several (4 – 5 in number) parallel chains of vortices with the distances Δy between them selected in such a way that the velocity profile $u_z(y)$ coincides with the wind velocity profile in the near-ground atmospheric layer ($-\infty < z < \infty$). Such an approach was used for modeling the periodic excitation of the two-dimensional turbulent mixing layer [56] as well as for simulating the interaction of an aircraft's vortex wake with a crosswind flow; in this case the vortex wake was represented as a two oppositely rotating vortex lines [71, 88]. For meeting the tangency conditions on the ground surface ($y = 0$), a system of mirror images of the vortex chains must be added at $y < 0$ with circulation of opposite sign.

The presence of a crosswind flow having the velocity invariant with height, as was indicated above (see, for example, Figs. 7.10–7.15), results in a flow asymmetry: both vortex tubes move horizontally (along the z axis) and vertically (along the y axis).

Chapter 8

AERODYNAMIC LOADS ON AIRCRAFT ENCOUNTERING VORTEX WAKES OF OTHER AIRCRAFT

8.1. Problem statement

Consider the motion of an aircraft of a specified configuration in disturbed air flow W, flying at subsonic speed V_k and arbitrary height H (Fig. 8.1).

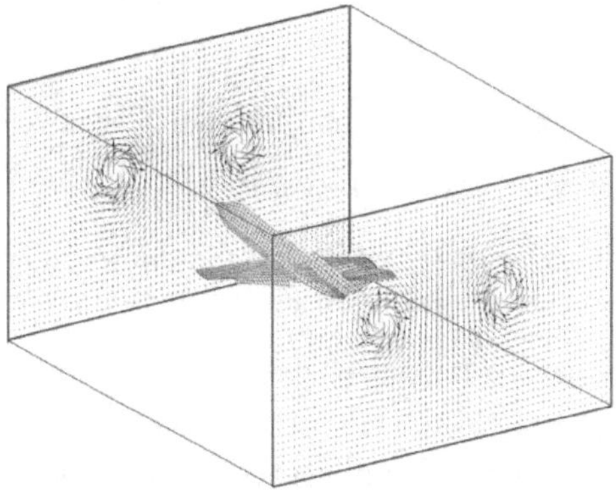

Fig. 8.1. The computational domain to determine aerodynamic loads on the aircraft

The following coordinate systems are used in this chapter: body axis system $O\,x\,y\,z$, air-path axis system $Ox_a\,y_a\,z_a$, and earth-fixed axis system $Ox_g\,y_g\,z_g$

The medium around the aircraft in hand is taken to be an ideal incompressible fluid, the aircraft's surfaces are assumed impermeable. The flow is potential everywhere outside the aircraft and its vortex wake. The vortex wake represents a thin vortex sheet, i.e., a surface with a jump of the tangential component of the velocity field. The flow separation lines are taken to be the sharp edges of the aircraft's surface.

The problem reduces to finding the velocity field $\vec{W}(M,t) = gradU\,(M,t)$ and pressure field $p\,(M,t)$ defined everywhere outside the wing surface σ_1 and the surface σ_2 representing the vortex wake. The vortex wake surface σ_2 is movable, its motion is not known in advance.

A.S. Ginevsky, A.I. Zhelannikov, *Vortex Wakes of Aircrafts*, Foundations of Engineering Mechanics, DOI 10.1007/978-3-642-01760-5_8, © Springer-Verlag Berlin Heidelberg 2009

Here, M(x, y, z) is a point in space, t is time. In calculating potential flows of an ideal incompressible fluid, we have for the velocity potential

$$\Delta U(Mt) = 0, \tag{8.1}$$

at all points of the computational domain outside the surfaces σ_1 and σ_2; for satisfying the equation of momentum it is sufficient to meet the Cauchy–Lagrange condition

$$\frac{P}{\rho} = \frac{P_\infty}{\rho} - \frac{W_\infty^2}{2} - \frac{\partial U}{\partial t}, \tag{8.2}$$

where P_∞ is the fluid pressure at infinity, ρ is the density of the fluid.

The flow tangency condition is satisfied on the aircraft surfaces:

$$\frac{\partial U^\pm}{\partial \overline{n}} = 0, \quad M \in \sigma_1, \tag{8.3}$$

where $\overline{n}$ is the unit vector normal to the surface σ_1 at the point M. On the vortex sheet at every moment of time the condition of zero pressure jump is met:

$$p^+(M,t) = p^-(M,t). \tag{8.4}$$

At infinity, the attenuation condition for disturbance velocities is satisfied:

$$\left| \nabla U(M,t) - \overrightarrow{W}_\infty \right| \to 0. \tag{8.5}$$

Thus, the statement of the problem at hand differs from the statements of the other problems of this book in that the motion of aircraft is considered in disturbed air flow, and the disturbances can be caused by another aircraft or other objects, including heights of a terrain.

8.2. A solution technique

The problem in hand is solved with the aid of the DVM in the framework of a nonlinear unsteady approximation, for which purpose the real aircraft is substituted with a schematized model. In this case the aircraft is treated as a combination of plate elements. In a number of works [8, 78, 83, 86] it was shown that such a representation is sufficient for obtaining acceptable results. Closed vortex frames were arranged on the model's plates. The schematized aircraft model was placed into a specified region of the vortex wake between two cross-flow planes, S_1 and S_2, for which disturbance velocities were precomputed (Fig. 8.2). The cross-flow planes were arranged in such a way that the aircraft under study was positioned within a rectangular parallelepiped between these planes.

With boundary conditions (8.1)–(8.5) met, the problem reduces to solving the system of linear algebraic equations

$$\sum_{i=1}^{n} \Gamma_i a_i = H_i - W_{ni}. \tag{8.6}$$

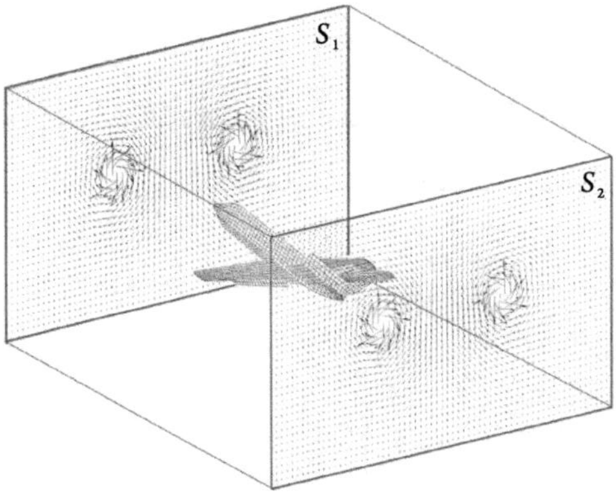

Fig. 8.2. The configurational domain between two cross-flow planes, $S1$ and $S2$, with precomputed flow parameters

The right sides of the equations, along with traditional functions H_i [7, 30], contain normal components of the disturbance velocities, W_{ni}, at the ith control point (the velocity W_{ni} is computed by linear interpolation between the disturbance velocities in the sections S_1 and S_2). Besides, the vortex sheet is also constructed with account for disturbance velocities. Aerodynamic loads are computed through the Cauchy–Lagrange integral (8.2), which takes into account, along with the undisturbed velocity, the disturbed flow velocity, too.

8.3. Verification of the method and predicted results

For checking the method for performance and the computed results for accuracy, computations were performed to be compared with available wind-tunnel experimental data.

Presented below as an example are the results of experiments carried out in the DNW wind-tunnel (the Netherlands) to study the effect of the vortex wake behind a vortex-generating wing on a following aircraft [23]. The configuration and dimensions of the encountering aircraft's model are given in Fig. 8.3.

Fig. 8.4 depicts the vortex-panel representations of the wake-generating wing and the encountering aircraft. The disturbance-velocity fields experimentally obtained with these models (Fig. 8.5) were compared with computed data (Fig. 8.6). A satisfactory agreement between theory and experiment can be observed.

The aircraft model was tested numerically and experimentally in the control plane with the disturbance velocity field shown in Fig. 8.5. The corresponding computed disturbance velocity field is presented in Fig. 8.6. The vortex wake behind the vortex-generating wing computed with the aid of nonlinear unsteady theory is depicted in Fig. 8.7. Using the computed

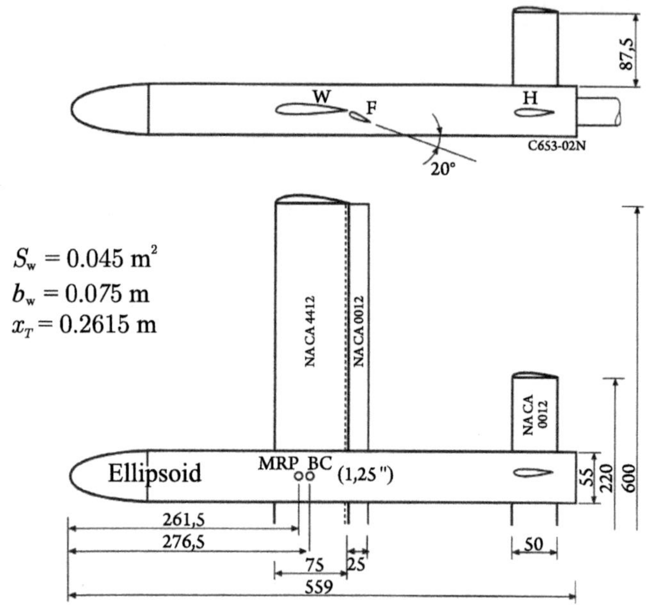

Fig. 8.3. Geometry and dimensions of the model representing a vortex-encountering aircraft

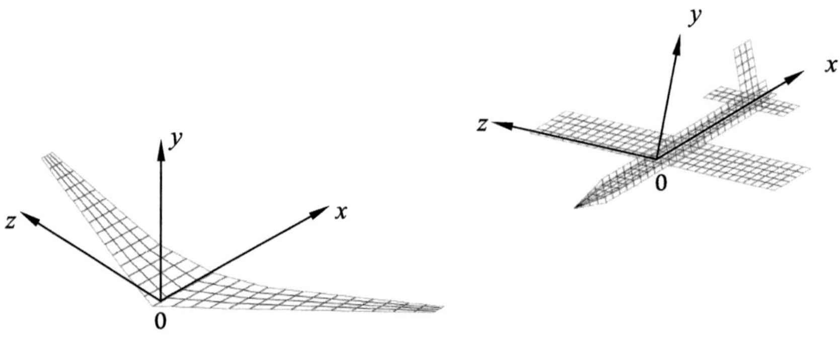

Fig. 8.4. Vortex-array representations of the vortex-generating wing and the encountering aircraft

velocity field, the aerodynamic loads on the encountering aircraft were obtained and compared with available experimental data (Fig. 8.8–8.11). Fig. 8.8 and 8.9 demonstrate the lift coefficient C_{ya} and the cross-stream force coefficient C_{za} as functions of the Y coordinate, whereas Figs. 8.10 and 8.11 show the dependence of the yawing and rolling moment

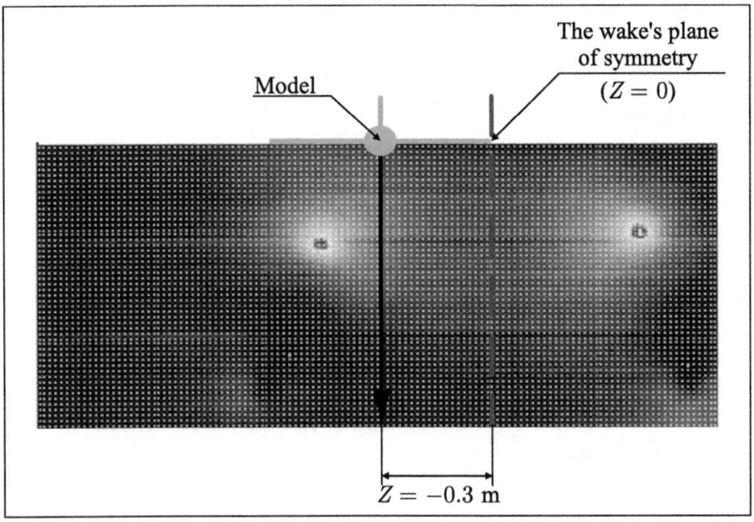

Fig. 8.5. Experimental disturbance-velocity field

Fig. 8.6. Computed disturbance-velocity field

coefficients on the same coordinate (solid line — experiment [72], line with dots — computation). Good agreement is observed between theory and experiment.

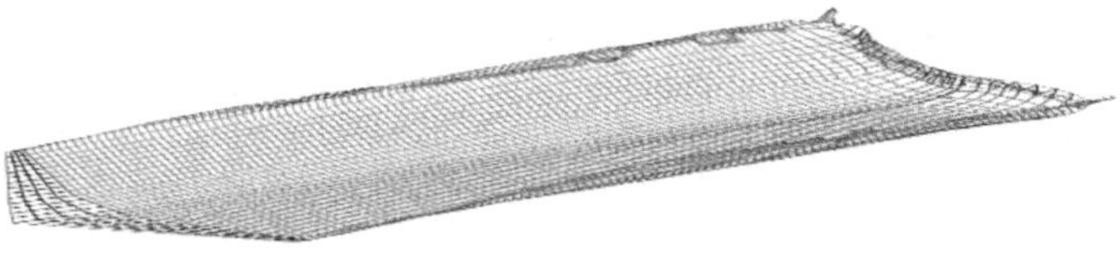

Fig. 8.7. Computed vortex wake behind the vortex-generating wing

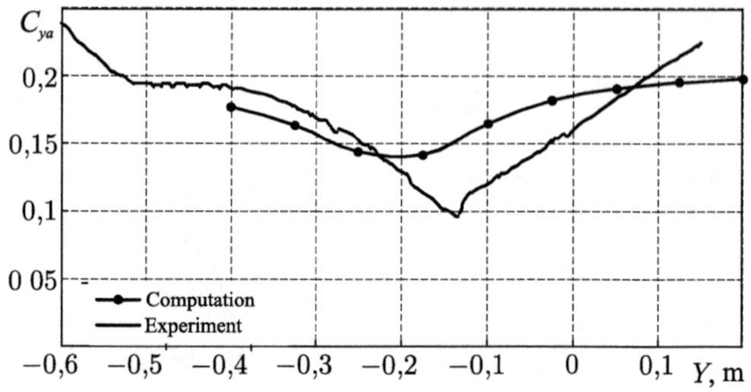

Fig. 8.8. C_{ya} versus Y coordinate

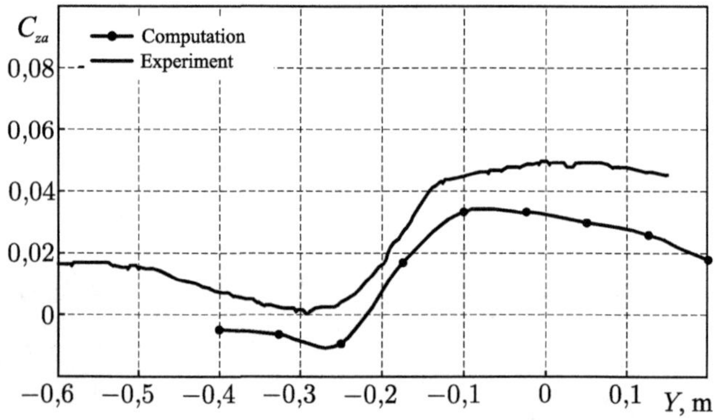

Fig. 8.9. C_{za} versus Y coordinate

8.4. The aerodynamic loads on aircraft in the far vortex wakes of preceding aircraft

Investigation of the loads on aircraft encountering vortex wakes of preceding aircraft shows potential wake-vortex hazards. Fig. 8.12 demonstrates the rolling moment coefficient of a Yak-40 aircraft along the $0z$ axis passing through the center of the right wing-tip vortex of a Il-76 vortex-generating aircraft. It can be seen that the values of the rolling moment coefficient reaches 0,06 in magnitude, which significantly exceeds the control authority m_x of the ailerons (available $m_x = 0{,}027$). Fig. 8.12 illustrates the location of the right wing-tip vortex 65 s after the passage of this section by the Il-76 at $H = 40\,\mathrm{m}$, $V = 340\,\mathrm{km/h}$; there was a crosswind of $W = -1\,\mathrm{m/s}$ from the right. The vortex is located at $H = 20\,\mathrm{m}$ over the runway centerline. For computation, the speed of the Yak-40 was taken to be $V = 230\,\mathrm{km/h}$.

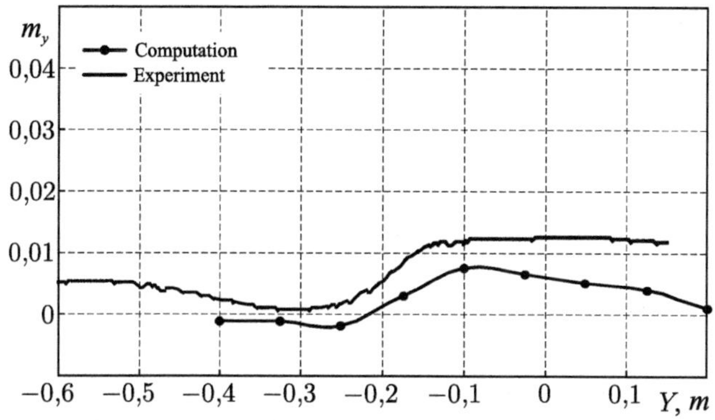

Fig. 8.10. Yawing moment coefficient m_y versus Y coordinate

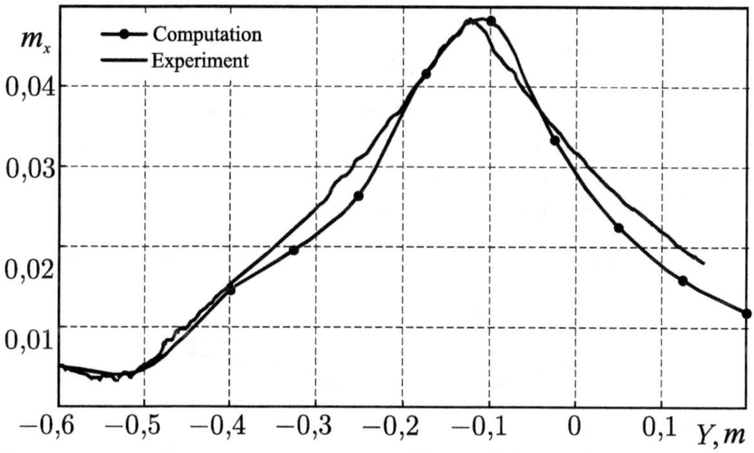

Fig. 8.11. Rolling moment coefficient m_x versus Y coordinate

Fig. 8.13 demonstrates the lift and pitching moment coefficients, C_{ya} and m_z, of the Yak-40 for the same flight conditions. The pitching moment coefficient is calculated with respect to the aircraft's center of mass. The local cross-stream force coefficient C_{za} and yawing moment coefficient m_y are presented in Fig. 8.14. One can see a significant variation of these coefficients Yak-40 along the $0z$ axis, whose origin coincides with the center of the right wing-tip vortex of the Il-76.

Fig. 8.15 demonstrates the variation of the rolling, yawing and pitching moment coefficients of the Yak-40 along the $0z$ axis passing through the center of the right wing-tip vortices of the Il-76, An-124, B-747 and A-380 aircraft. All vortex-generating aircraft had the same flight speed, $V = 300$ km/h, and flight height, $H = 40$ m. The Yak-40 entered the vortex wake after 25 s following the passage of each of these

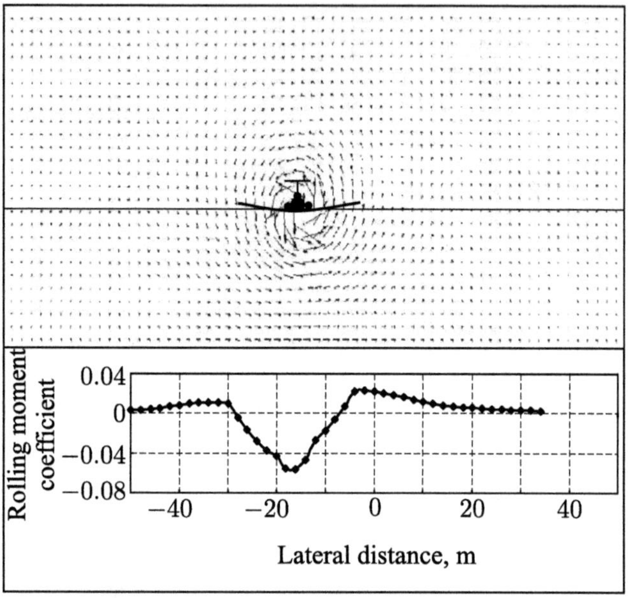

Fig. 8.12. Variation of the rolling moment coefficient mx with the lateral distance for the Yak-40 in the vortex wake of the Il-76, $t = 65\,\mathrm{s}$

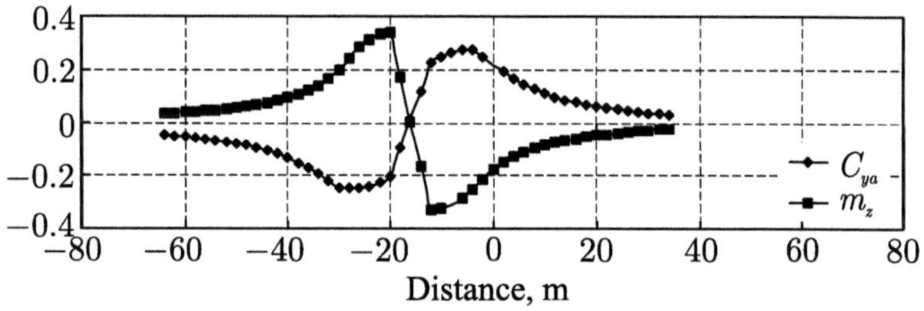

Fig. 8.13. Variation of the lift and pitching moment coefficients, C_{ya} and m_z with the lateral distance for the Yak-40 in the vortex wake of the $t = 65\,\mathrm{s}$

vortex-generating aircraft through this control plane, i.e., the distance between the Yak-40 and the preceding aircraft was $X = 2{,}1$ km. The atmospheric conditions were taken to be stable (CN= 1). One can see that under these conditions the vortex wakes pose a serious hazard to the Yak-40, especially in roll. For example, the A-380 vortex wake induces the rolling moment coefficient equal to $m_x = -0{,}2$, which is on the order of magnitude greater than the Yak-40 aileron authority equal to $m_x = -0{,}027$.

Fig. 8.16 presents the dependence of the lift and cross-stream force coefficients of the Yak-40 along the $0z$ axis passing through the center of the right wing-tip vortices of the Il-76, An-124, B-747 and A-380 aircraft.

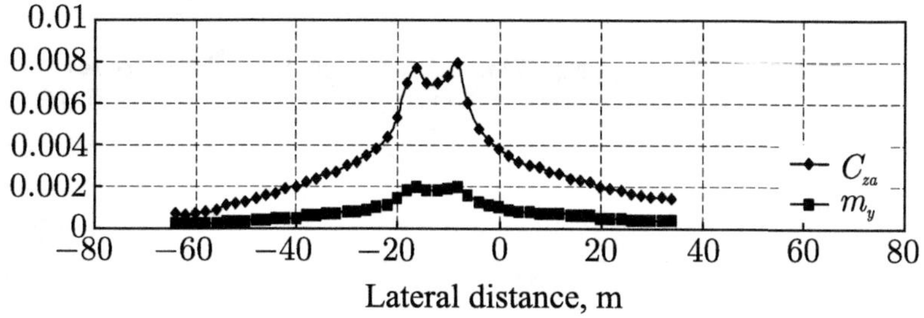

Fig. 8.14. Variation of the cross-stream force coefficient C_{za} and yawing moment coefficient m_y with the lateral distance for the Yak-40 in the vortex wake of the Il-76, $t = 65$ s

All vortex-generating aircraft had the same flight speed, $V = 300$ km/h, their flight height was $H = 40$ m. The Yak-40 entered the vortex wake also after 25 s following the passage of the control plane by a vortex-generating aircraft. To illustrate the loads imposed upon the Yak-40 in vortex wakes, Fig. 8.17 demonstrates the disturbance-velocity fields generated by the Il-76, An-124, B-747 and A-380 at $t = 25{,}2$ s at a distance of 2,1 km behind them. The scale division is 10 m/s.

8.5. Prediction of the effect of wind flow over rough terrain on the aerodynamic loads experienced by an aircraft

Investigation into an aircraft's behavior in wind flows affected by a rugged terrain shows that such flows can pose a hazard for the aircraft. As an example, let us consider the computed behavior of a Su-25 fighter aircraft flying over a mountainous terrain. Fig. 8.18 demonstrates its trajectory over a mountain group featuring a developed vortex structure corresponding to that depicted in Fig. 6.3. Consider the flight conditions at three points on the trajectory at heights of $H = 125$ m, 250 m and 375 m. Using the methodology of Section 8.2, the aircraft's aerodynamic characteristics and its vortex wake are then computed. An example of the computation of the Su-25's vortex wake in a terrain-affected wind flow is given in Fig. 8.19.

Shown in Fig. 8.20 is the Su-25 aircraft's lift coefficient C_{ya} along the trajectory shown in Fig. 8.18 at heights of $H = 125$ m, 250 m and 375 m. It can be seen that under these conditions the strongest effect of the wind flow on the aircraft takes place at $H = 250$ m. Fig. 8.21 demonstrates the variation of the cross-stream force coefficient C_{za} along the trajectory at the same heights. The largest changes in the cross-stream force coefficient correspond to $H = 250$ m.

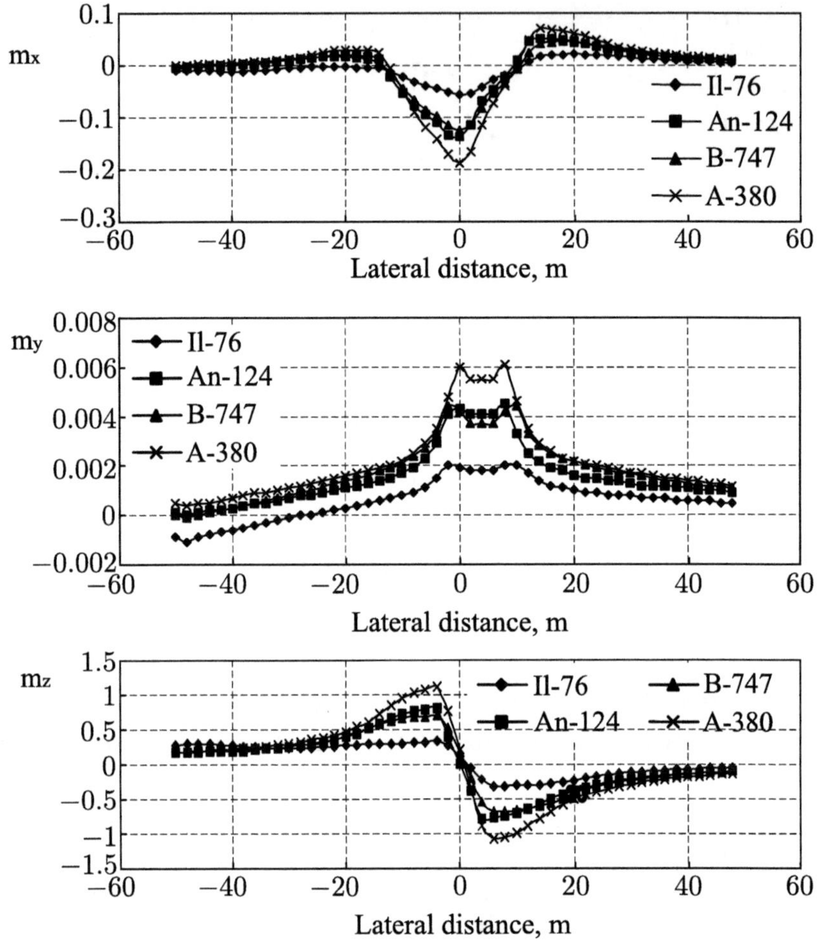

Fig. 8.15. Variation of the rolling (mx), yawing (my), and pitching (mz) moment coefficients of the Yak-40 with the 0z coordinate in the vortex wake behind the Il-76, An-124, B-747 and A-380.

The variations of the rolling, yawing and pitching moment coefficients, m_x, m_y and m_z respectively, along the indicated trajectory at $H = 125$ m, 250 m and 375 m are shown in Figs. 8.22–8.24.

In Figs. 8.22–8.24, the thin dashed lines show the trim pitching moment coefficient; dash-dotted lines mean the values of the moment coefficients which can be counteracted with a 25 percent deflection of the corresponding control surfaces. The thick dashed lines show the values of moment coefficients which require a 50 percent deflection of the control surfaces to be counteracted. One can see that for the flight conditions

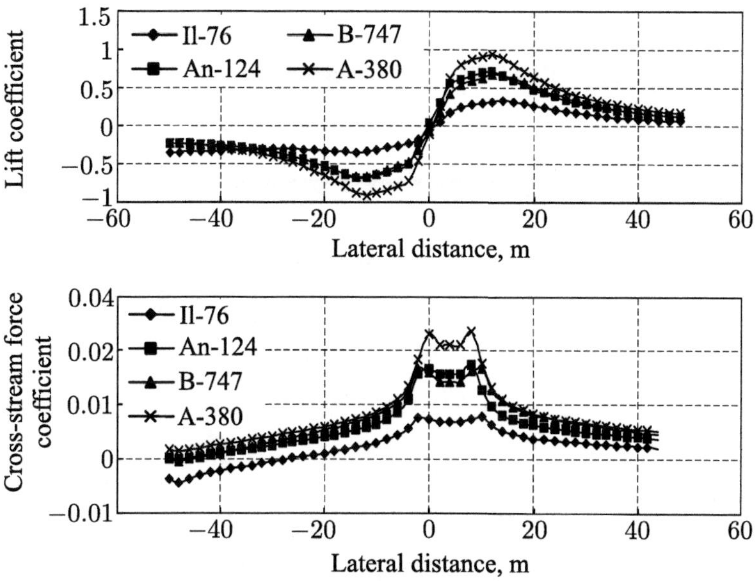

Fig. 8.16. Variation of the lift and cross-stream force coefficients of the Yak-40 with the $0z$ coordinate in the vortex wake of the Il-76, An-124, B-747 and A-380

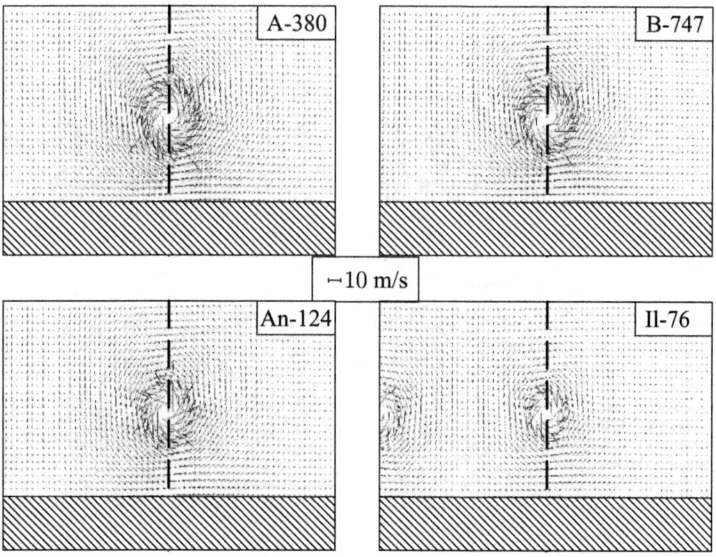

Fig. 8.17. The right wing-tip vortex behind the Il-76, An-124, B-747 and A-380 aircraft; $t = 25{,}2$ s; downstream distance 2,1 km

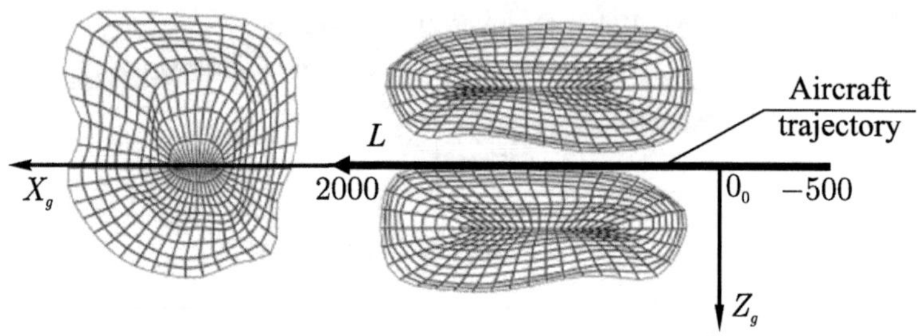

Fig. 8.18. The flight trajectory of the Su-25 aircraft

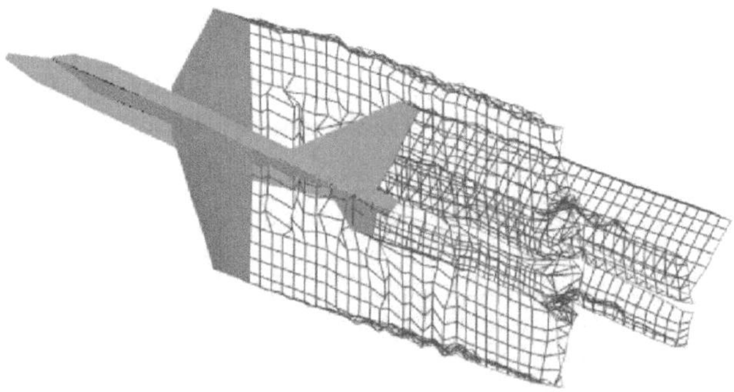

Fig. 8.19. Computed vortex wake of the Su-25 aircraft with account for the effect of wind flow over terrain

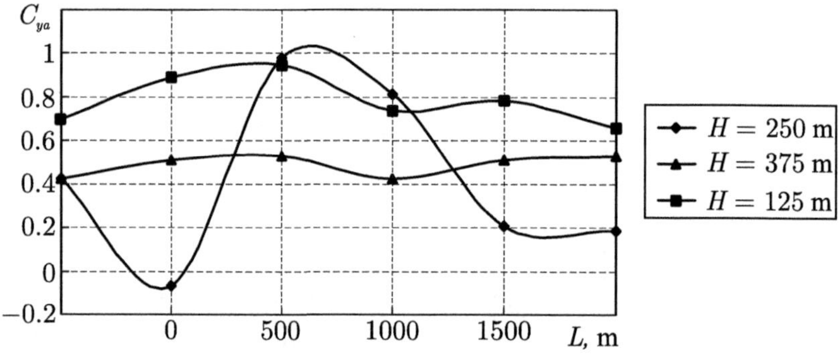

Fig. 8.20. Variation in lift coefficient along the flight trajectory at different heights

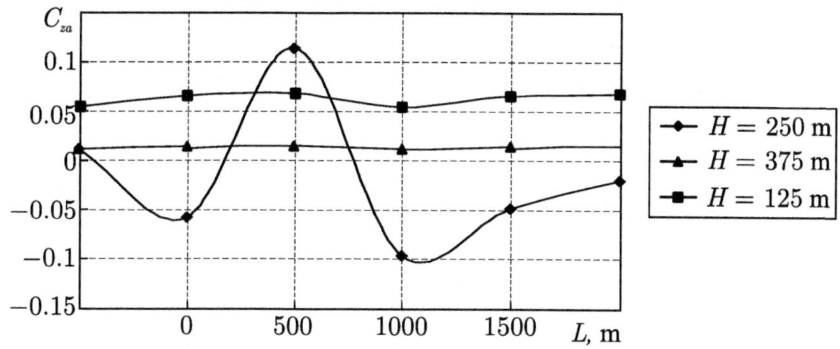

Fig. 8.21. Variation in cross-stream force coefficient along the flight trajectory at different heights

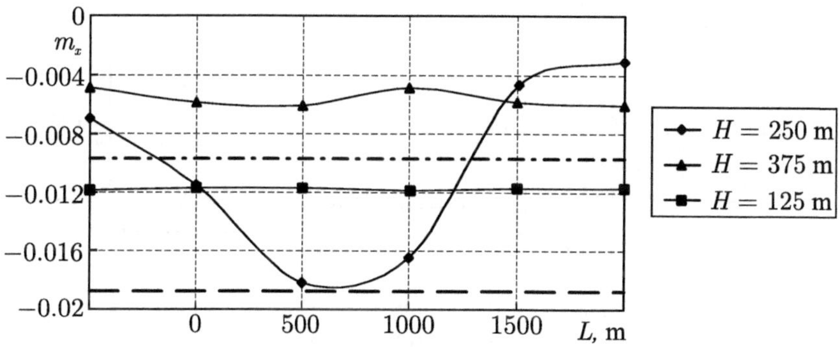

Fig. 8.22. Rolling moment coefficient variation along the flight trajectory at different flight heights

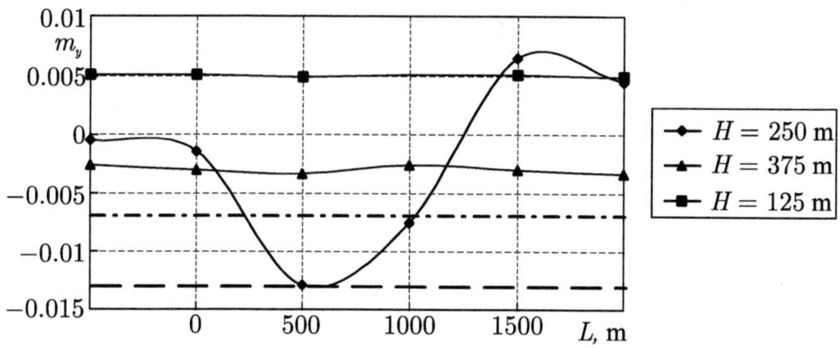

Fig. 8.23. Yawing moment coefficient variation along the flight trajectory at different flight heights

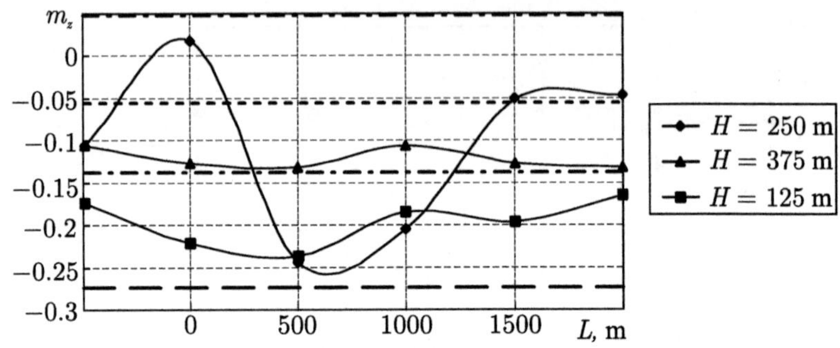

Fig. 8.24. Pitching moment coefficient variation along the trajectory at different flight heights

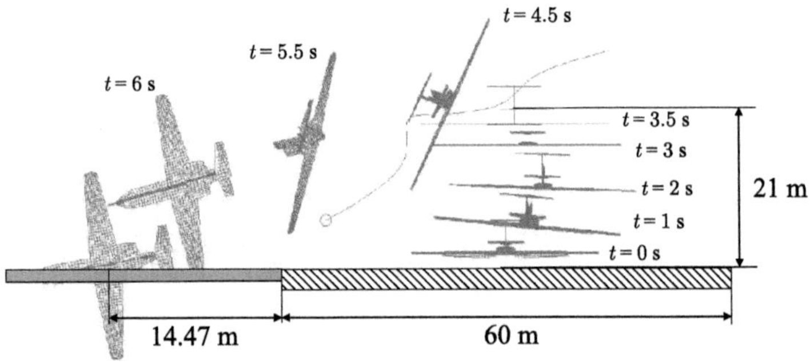

Fig. 8.25. Computed motion of the Yak-40 aircraft in the vortex wake of the Il-76 aircraft (rear view)

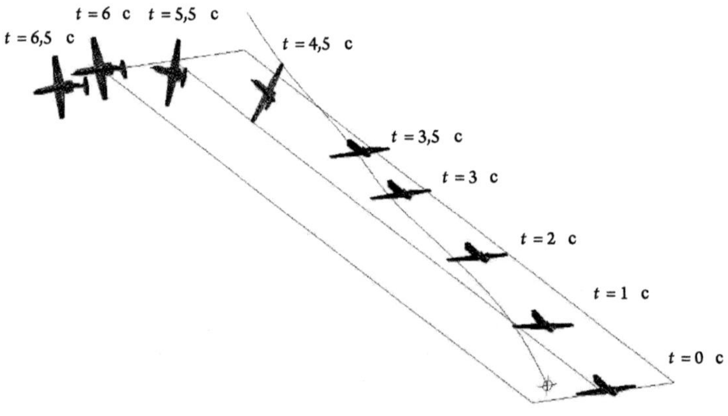

Fig. 8.26. The trajectory of the Yak-40 in the Il-76's vortex wake (3/4 left rear view)

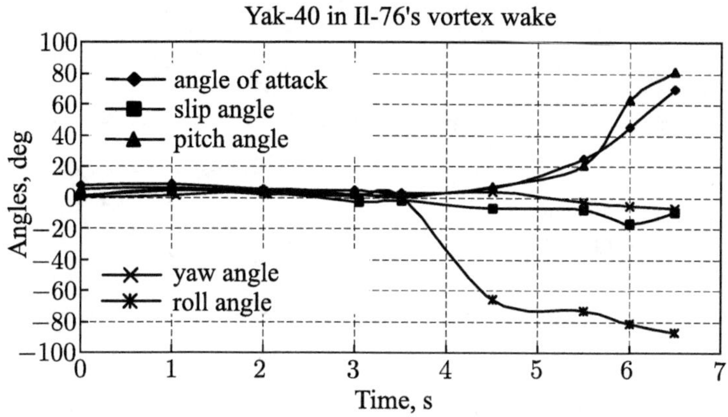

Fig. 8.27. Time histories of the Yak-40's angle of attack and slip, pitch, yaw and roll angles

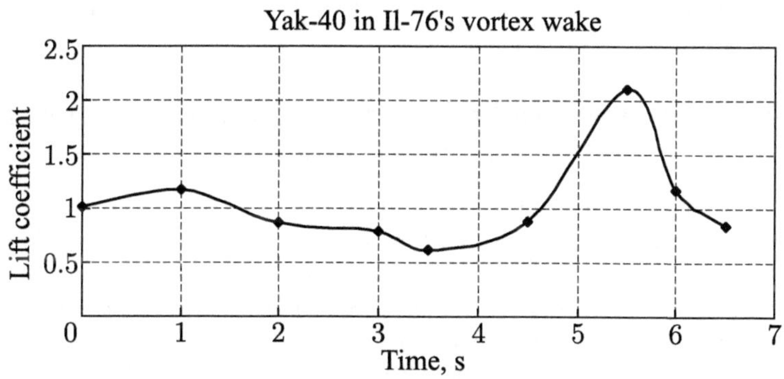

Fig. 8.28. Time history of the lift coefficient of the Yak-40

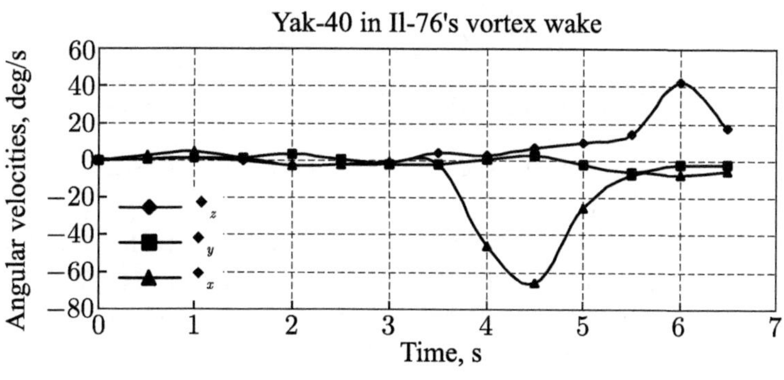

Fig. 8.29. Time histories of the Yak-40's angular velocities ω_x, ω_y and ω_z

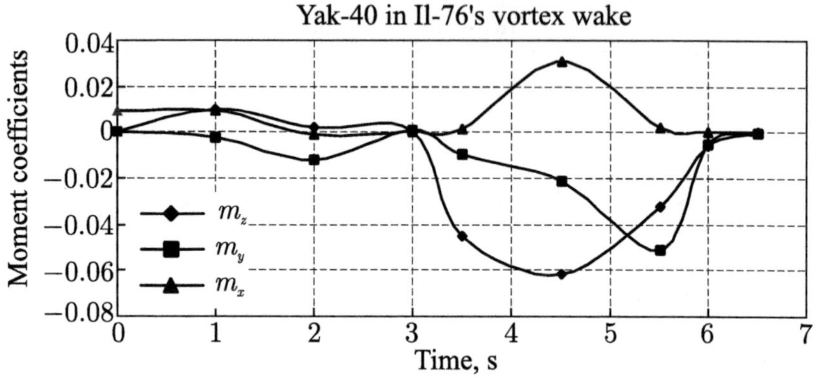

Fig. 8.30. Time histories of the rolling, yawing and pitching moment coefficients (m_x, m_y m_z respectively).

considered at $H = 250$ m the moments acted on the aircraft Su-25 may require 50% of the full control surface travel to trim the aircraft.

8.6. Numerical prediction of an aircraft's dynamics in a vortex wake

For predicting the motion of an aircraft under the action of forces and moments due to encountering with a vortex wake, the equations of motion of the rigid body will be used in the body axis coordinate system. The aircraft's mass and moments of inertia are taken to be constant within the time intervals considered and equal to the initial ones. With the axes of the body-axis system coinciding with the main axes of inertia, the system of equations governing the motion of an aircraft about its center of mass takes the form:

$$\begin{aligned} m\left(\frac{dV_x}{dt} + \Omega_y V_z - \Omega_z V_y\right) &= X + P_x - G\sin\vartheta, \\ m\left(\frac{dV_y}{dt} + \Omega_z V_x - \Omega_x V_z\right) &= Y + P_y - G\cos\vartheta\cos\gamma, \\ m\left(\frac{dV_z}{dt} + \Omega_x V_y - \Omega_y V_x\right) &= Z + P_z + G\cos\vartheta\sin\gamma, \end{aligned} \tag{8.7}$$

where V_x, V_y, V_z are the components of the velocity vector $\vec{V}_k$; Ω_x, Ω_y, Ω_z are the components of the angular velocity vector $\vec{\Omega}$; X, Y, Z are the components of the aerodynamic force $\vec{R}_{textA}(t)$; P_x, P_y, P_z are the components of the engine thrust $\vec{P}(t)$.

Under the assumptions made, the angular motion of the aircraft is described by the Euler equations:

$$
\begin{aligned}
I_x \frac{d\Omega_x}{dt} + (I_z - I_y)\,\Omega_y\Omega_z &= M_x + M_{Px} + M_{\Gamma x},\\
I_y \frac{d\Omega_y}{dt} + (I_x - I_z)\,\Omega_x\Omega_z &= M_y + M_{Py} + M_{\Gamma y},\\
I_z \frac{d\Omega_z}{dt} + (I_y - I_x)\,\Omega_x\Omega_y &= M_z + M_{Pz} + M_{\Gamma z},
\end{aligned}
\tag{8.8}
$$

where I_x, I_y, I_z are the moments of inertia; M_x, M_y, M_z, M_{Px}, M_{Py}, M_{Pz}, $M_{\Gamma x}$, $M_{\Gamma y}$, $M_{\Gamma z}$ are the components of the moments $\vec{M}(t)$, $\vec{M}_P(t)$ и $\vec{M}_\Gamma(t)$, respectively.

Systems of equations (8.7) and (8.8) allow one to find the components of the translational and angular velocities. To determine the coordinates of the aircraft's center of mass and the position of the body-fixed coordinate system relative to the earth-fixed axis system it is necessary to use the six kinematic differential equations.

$$
\begin{aligned}
\frac{d\vartheta}{dt} &= \Omega_y \sin\gamma + \Omega_z \cos\gamma,\\
\frac{d\gamma}{dt} &= \Omega_x - tg\vartheta\,(\Omega_y \cos\gamma - \Omega_z \sin\gamma),\\
\frac{d\psi}{dt} &= \frac{\Omega_y \cos\gamma - \Omega_z \sin\gamma}{\cos\vartheta},\\
\frac{dx_0}{dt} &= V_x \cos\vartheta\cos\psi + V_y(\sin\gamma\sin\psi - \cos\gamma\cos\psi\sin\vartheta) +\\
&\qquad + V_z(\cos\gamma\sin\psi + \sin\vartheta\sin\gamma\cos\psi),\\
\frac{dy_0}{dt} &= V_x \sin\vartheta + V_y\cos\gamma\cos\vartheta - V_z\sin\gamma\cos\vartheta,\\
\frac{dz_0}{dt} &= -V_x\cos\vartheta\sin\psi + V_y(\sin\gamma\cos\psi + \cos\gamma\sin\vartheta\sin\psi) +\\
&\qquad + V_z(\cos\gamma\cos\psi - \sin\gamma\sin\vartheta\sin\psi).
\end{aligned}
\tag{8.9}
$$

The systems of equations (8.7) and (8.9) with specified initial conditions uniquely determine the motion of aircraft. However, for solving equations (8.7) and (8.8) it is necessary to know the aerodynamic force $\vec{R}_A(t)$ and moment $\vec{M}(t)$, acting on the aircraft (with the proviso that all other quantities, $\vec{P}(t)$, $\vec{M}_P(t)$ and $\vec{M}_\Gamma(t)$ are known). Thus, for solving the systems of equations (8.7)–(8.9) using the method described in Section 8.2 we shall determine at each instant of time the aerodynamic force $\vec{R}_A(t)$ and aerodynamic moment $\vec{M}(t)$ experienced by the aircraft.

If necessary, the actions of the pilot also can be taken into consideration. Control surface deflections are governed by the aircraft's control laws, dynamic properties of the control system and the pilot. For modeling the aircraft's control system, let us make the following assumptions:

— control linkage is completely rigid;

— the linkage with distributed mass and stiffness properties is replaced by a lumped-parameter model;

— no looseness and dry friction in the linkage;

— viscous friction forces are taken to be proportional to the control stick and surfaces' deflection rate;

— load feel mechanism's stiffness is constant.

With the above assumptions, the control system's elements at the pilot's $f = 1 - 3\,\text{Hz}$ operating frequencies can be considered as amplifying. In this case the transfer function of the control system takes the form:

$$W_{\text{CS}} = \frac{K_{\text{air}}}{C_{\text{air}}}, \tag{8.10}$$

where K_{air} is the control linkage gear ratio (from the control stick to the control surface); C_{air} is the stick's feel mechanism stiffness.

The pilot is the most complex element of the control loop. His performance varies within wide limits in accordance with his physical states influenced by a multitude of factors. However, the following general properties can be singled out:

— the pilot generates commands as a function of the magnitude and direction of an external disturbance or the magnitude of an error signal;

— the pilot's response to an external disturbance appears not at once but after a certain delay which depends on the direction and intensity of the disturbance, the number of parameters under the pilot's control and his psychophysiological state. To an intensive disturbing acceleration the pilot responds reflectively with a delay of 0.13–0.6 s;

— the pilot has a zone of insensibility (dead zone). If the disturbance intensity is below the sensitivity threshold, the pilot does not respond to it;

— being aware of his delay, the pilot tries to operate with a lead.

A single common model suitable for the pilot's actions in all cases does not exist. For each concrete problem a special model, discrete or continuous, is usually developed. To control or stabilize a parameter x_i the following transfer function for the pilot's actions is specified:

$$W_{\text{p}}(P) = W_{\frac{\Delta P}{\Delta x_i}}(P) = \frac{K_{\text{p}} e^{-\tau P}\left(T_{\text{p}1} P + 1\right)}{\left(T_{\text{p}2} P + 1\right)\left(T_{\text{p}3} P + 1\right)}, \tag{8.11}$$

where $\Delta x_i = x_i - x_{i,\,specified}$; $K_{\text{p}} = \dfrac{\Delta P}{\Delta x_i}$ is the pilot's gain, $\tau = 0{,}13 \ldots 0{,}3\,\text{s}$ is the pilot's delay; $\dfrac{1}{T_{\text{p}2}P + 1}$ is the element to account for the pilot's inertia in receipt and recognition of information; $T_{\text{p}2} \leqslant 2\,\text{s}$; $\dfrac{1}{T_{\text{p}3}P + 1}$ is the element accounting for the neuromuscular delay of the pilot; $T_{\text{p}3} = 0{,}1 \ldots 0{,}3\,\text{s}$; $T_{\text{p}1}P + 1$ accounts for the pilot's ability to allow a lead, $T_{\text{p}1} \leqslant 2{,}5\,\text{s}$.

Thus, the nonlinear mathematical model of the aircraft's 3D dynamics with allowance for the pilot's control actions is described in the general case with the system of equations (8.7)–(8.11). As seen from these equations, there is a complex interrelation between the parameters involved.

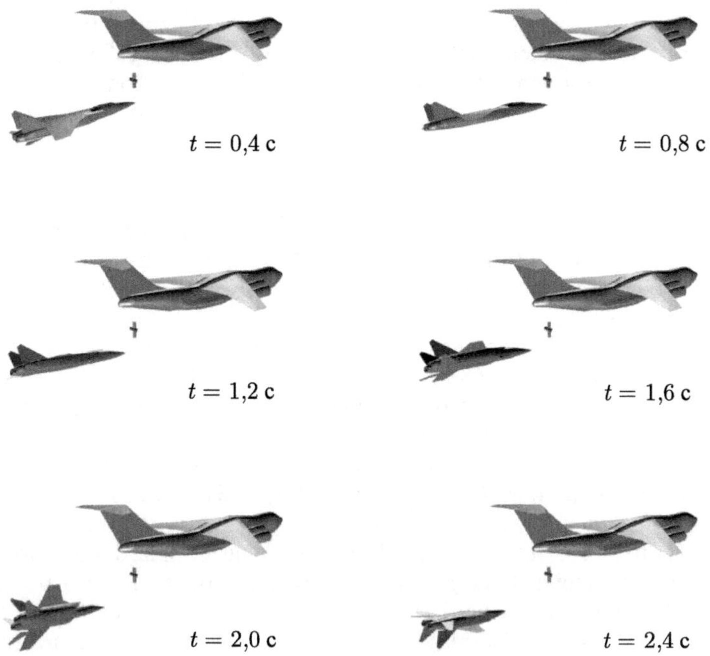

Fig. 8.31. The process of aerial refueling without the pilot's intervention to counteract the upset caused by the tanker's vortex wake

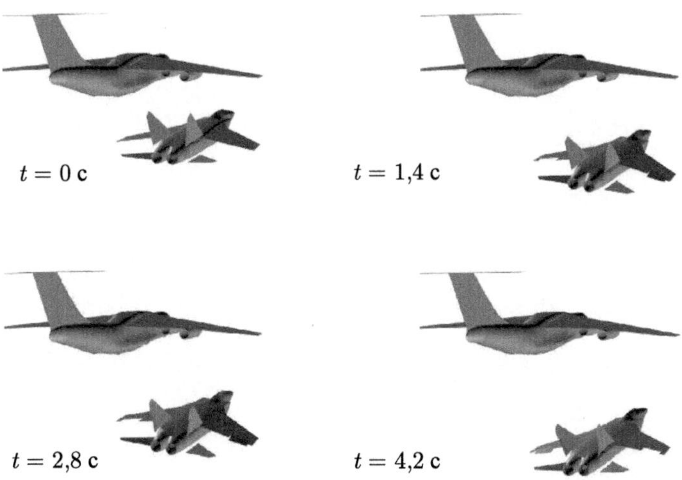

Fig. 8.32. The process of aerial refueling with proper intervention of the pilot of the refueled aircraft to flight control to counteract the upset caused by the trailing wake

So, the motion of aircraft depends on aerodynamic loads acting on them. In turn, the values of aerodynamic forces and moments depend on the aircraft's motion characteristics. The control surfaces' deflections determine the aerodynamic loads and the aircraft's motion. Because of this, it is necessary to simultaneously solve the equations of flight dynamics, nonlinear unsteady aerodynamics and those describing the control loop. In practice these equations are solved numerically in a step-by-step manner in time.

As an example of computation through this algorithm, Figs. 8.25 and 8.26 present the motion of the Yak-40 aircraft in the vortex wake of the Il-76, reconstructing the Yak-40's 16 January 1987 crash in Tashkent (see Fig. 4.8).

Figs. 8.25–8.30 demonstrate in detail the time histories of various quantities (angle of attack, pitch, slip, yaw and roll angles, lift coefficient, angular velocities and moment coefficients) characterizing the motion of the Yak-40 during the crash.

The presented methodology can be used to predict aircraft's dynamics during air refueling. As an example, Fig. 8.31 demonstrates the position of a refueled aircraft (MiG-31) relative to a tanker aircraft (Il-78). The refueling is carried out from the right-hand standard hose-drogue unit. The figure shows both aircraft at $t = 0{,}4$ s; 0,8 s; 1,2 s; 1,6 s; 2,0 s and 2,4 s for the case without the pilot's intervention to counteract the upset caused the tanker aircraft's vortex wake. One can see that ultimately the receiver aircraft becomes rolled upside down. The proper flight control actions by the pilot ensure normal aerial refueling (see Fig. 8.32).

References

1. *Airapetov AB* (1977) On the statistical properties of a discrete vortex system modeling turbulent jet flow (in Rusian). In: Turbulent flows, Nauka, Moscow, pp. 188-195
2. *Anashkin AG, Bondarenko VM, Zhelannikov AI* (1989) A study into the aerodynamic characteristics of Yak-40 aircraft in the far aerodynamic wake behind an Il-76 (in Russian). In: Scientific methodological material on aircraft aerodynamics. Zhukovsky Air Force Engineering Academy, Moscow
3. *Aparinov VA* (1977) Prediction of the nonlinear aerodynamic characteristics of a wing with complex planform with account for the leading edge vortex sheet (in Russian). Izv. AN USSR, Mech Fluid Gas № 2
4. *Aparinov VA* (1986) Computer-based modeling of complex 3D motion of lifting surfaces with a developing unsteady vortex wake (in Russian). In: Computer-aided investigation of aircraft aerodynamic characteristics. Proc Zhukovsky Air Force Engineering Academy, Moscow, № 1313: 200 – 222
5. *Aparinov VA, Belotserkovsky SM, Lifanov IK, Mikhailov AA* (1987) Computation of unsteady aerodynamic characteristics of bodies with separated flow about them (in Russian). J Comp Math Math Phys 28, 1558 – 1566
6. *Aparinov VA, Dvorak AV* (1986) Discrete vortex method with closed vortex frames (in Russian). In: Computer-aided investigation of aircraft aerodynamic characteristics. Proc Zhukovsky Air Force Academy, Moscow № 1313, 424 – 432
7. *Aubakirov TO, Belotserkovsky SM, Zhelannikov AI, Nisht MI* (1997) Nonlinear wing theory and its applications (in Russian). Almaty, Gylym, 448 pp.
8. *Aubakirov TO, Ivanov PE, Zhelannikov AI, Nisht MI* (1999) Trailing wakes and their effect on aircraft: computer simulation (in Russian). Almaty, 230 pp.
9. *Babkin VI, Belotserkovsky AS, Turchak et al* (2008) A system for providing vortex-related flight safety for aircraft (in Russian). Nauka, Moscow, 374 pp.
10. *Baranov NA, Belotserkovsky SM, Belotserkovsky AS, Ginevsky AS* (1997) Computation of eddy viscosity in the trailing vortex wake behind an aircraft (in Russian). Preprint TsAGI № 109, 16 pp.
11. *Belotserkovskii OM* (1999) Turbulence and instabilities. Moscow Institute of Physics and Technology, 348 pp
12. *Belotserkovskii SM* (1991) Computerized study of vortex aerohydrodynamics of aircraft and helicopters. Proc. Aircraft Wake Vortices Conf, Washington, 1, 20-1 – 20-33
13. *Belotserkovskii SM* (1991) Wake vortex research in the USSR. Proc Aircraft Wake Vortices Conf, Washington, 1, 7-1 – 7-25

A.S. Ginevsky, A.I. Zhelannikov, *Vortex Wakes of Aircrafts*, Foundations of Engineering Mechanics,
DOI 10.1007/978-3-642-01760-5_BM2, © Springer-Verlag Berlin Heidelberg 2009

14. *Belotserkovsky AlS, Ginevsky AS* (2002) Numerical simulation of the far vortex wake during takeoff and landing operations (in Russian). Russ Acad Sci Reports 380, № 6: 761 – 764

15. *Belotserkovsky AlS, Ginevsky AS* (2003) Evolution of the far vortex wake of an aircraft near the ground surface with account for atmospheric turbulence (in Russian). Proc TsAGI № 2643: 9 – 27

16. *Belotserkovsky AlS, Ginevsky AS, Pogrebnaya TV, Shipilov SD* (2004) Numerical visualization of the far vortex wake of aircraft at takeoff and landing (in Russian). Aeronautical Engineering 77, № 1: 32 – 36

17. *Belotserkovsky AlS, Ginevsky AS, Pogrebnaya TV, Shipilov SD* (2000) Simulation of the far vortex wakes of the Tu-204 and Il-96 trunk-route aircraft during takeoff and landing operations (in Russian). Aeronautical Engineering 76, № 6: 12 – 20

18. *Belotserkovsky AlS, Ginevsky AS, Pogrebnaya TV, Shipilov SD* (2003) Numerical simulation of the far vortex wake of trunk-route aircraft during takeoff and landing operations (in Russian). Mech Uspekhi 2, № 4: 106 – 127

19. *Belotserkovsky AS, Pasekunov IV* (2005) An integrated system for ensuring vortex-related flight safety (in Russian). Aerospace Review № 1

20. *Belotserkovsky AS, Zhuritsky GI, Kanevsky MI, Pasekunov IV* (2007) Vortex-related flight safety – the real state of the art (in Russian). The World's Air Shows № 3: 40 – 45

21. *Belotserkovsky OM* (1994) Numerical simulation in continuous media mechanics (in Russian). Fizmatlit, Moscow, 442 pp.

22. *Belotserkovsky SM, Dvorak AV, Zhelannikov AI, Kotovsky VN* (1987) Computer-based simulation of turbulent jets and wakes. In: Problems of turbulent flows (in Russian). Nauka, Moscow

23. *Belotserkovsky SM, Ginevsky AS* (1995) Computer-based concept of vortex turbulence (in Russian). In: Applied and nonlinear dynamics. Proc Institutes of Higher Education 3, № 2: 72 – 93

24. *Belotserkovsky SM, Ginevsky AS* (1995) Simulation of turbulent jets and wakes using the method of discrete vortices (in Russian). Fizmatlit, Moscow, 368 pp.

25. *Belotserkovsky SM, Ginevsky AS, Khlapov NV* (1995) Simulation of a round turbulent jet using the method of discrete vortices (in Russian). Russ Acad Sci Reports 345, № 4: 479 – 482

26. *Belotserkovsky SM, Kotovsky VN, Nisht MI, Fedorov RM* (1988) Mathematical simulation of separated flows about bodies (in Russian). Nauka, Moscow, 232 pp.

27. *Belotserkovsky SM, Lifanov IK* (1985) Numerical methods in singular integral equations and their application to aerodynamics, theory of elasticity and electrodynamics (in Russian). Nauka, Moscow

28. *Belotserkovsky SM, Loktev BE, Nisht MI* (1992) Computed-aided investigation of aerodynamic and aeroelastic characteristics of helicopter rotors (in Russian). Mashinostroenie, Moscow

29. *Belotserkovsky SM, Medvedev RE, Pogrebnaya TV, Shipilov SD* (2001) Simulation of the near vortex wake of an aircraft with deployed sectionalized flaps (in Russian). Proc TsAGI № 2643: 28 – 35

30. *Belotserkovsky SM, Nisht MI* (1978) Separated and attached flows of an ideal fluid past thin wings (in Russian). Nauka, Moscow

31. *Belotserkovsky SM, Shipilov SD, Pogrebnaya TV* (1995) Computer-aided investigation into aircraft's aerodynamic derivatives (in Russian). Russ Acad Sci Reports 341

32. *Bernard PS* (2006) Turbulent flow properties of large-scale vortex systems. Proc Nat Acad Sci USA 103, № 276: 10174 – 10179

33. *Bushgens GS* (Ed.) (1995) Aerodynamics and flight dynamics of trunk-route aircraft (in Russian). TsAGI Publishing Division and Aviation Publishing House of People's Republic of China, Moscow-Beijing, 772 pp.

34. *Byzova IL, Garger EK, Ivanov VN* (1991) Experimental studies of atmoshere diffusion and computation of admixture spreading (in Russian). Gidrometeoizdat, Moscow, 280 pp.

35. *Campbell SD, Dasay TJ, Freehart RE et al* (1996) Wake vortex field measurement program of Memphis. TN Data Guine, Lincoln Laboratory, Lexington, Massachusetts, Project Report ATS-240, 93 pp.

36. *Corjon AQ, Risso F, Stoessel F*, Poinsot T (1996) Tree-dimensional direct numerical simulations of wake vortices: atmosphere turbulence effect and rebound with crosswind. AGARD Conf Proc, 584, 28-1 – 28-21

37. *Darracq D, Corjon A, Dueros F* (2000) Simulation of wake vortex detection of airborne Doppler lidar. J Aircraft 37, № 6

38. *Dee JK, Nicholas OP* (1968) Flight measurement of wing tip vortex motion near the ground. Techn Rep, 68007, RAE

39. *Fedyaevsky KK, Ginevsky AS, Kolesnikov AV* (1973) Computation of the turbulent boundary layer of an incompressible fluid (in Russian). Sudostroyenie, Leningrad, 256 pp.

40. *Ferreira Gago, Brunet S, Garnier F* (2002) Numerical investigation of turbulent mixing in a jet/wake vortex interaction. AIAA J 40, 276 – 285

41. *Gaifullin AM* (2006) Investigation of vortex structures generated in gas or fluid flows past bodies (in Russian). TsAGI, Moscow, 138 pp.

42. *Gaifullin AM, Zvonova YuS, Sviridenko YuN* (2002) Prediction of interference between a turbulent exhaust jet and the aircraft airframe (in Russian). Proc TsAGI № 2655: 160 – 166

43. *Ginevsky AS, Pogrebnaya TV, Shipilov SD* (2007) Modeling of motion of a vortex pair near the ground board using the discrete vortex method (in Russian). In: Methods of discrete singularities in problems of mathematical physics. Int Symp Proc, Kharkov, pp. 105 – 109

44. *Golovnev AV, Gulyaev VV, Ikryannikov ED, Maslov VP, Smelter VV, Rakhmatullin RSh* (2005) The use of the discrete vortex method in solving the problems of aerodynamic design of aviation weaponry (in Russian). Vooruzheniye, Moscow

45. *Golovnev AV, Zhelannikov AI, Eremenko SM, Nekrakha SI* (2005) Account for the ground effect in computing unsteady flow around aircraft (in Russian). Material of the XVI school-symposium «Aerodynamics of aircraft», TsAGI
46. *Greene G* (1986) An approximate model of vortex decay in the atmosphere. J Aircraft 23, 566 – 573
47. *Grinats ES, Kashevarov AV, Stasenko AL* (1996) Numerical investigation of the jet-vortex wake of a heavy aircraft near the ground (in Russian). Proc TsAGI № 2622: 137 –145
48. *Harvey JK, Perry FT* (1971) Flow field produced by trailing vortices in the vicinity of the ground. AIAA J 9, 1659 – 1661
49. *Jacquin L, Garnier F* (1995) On the dynamics of engine jets behind a transport aircraft. AGARD-CP-584, 37-1 – 37-8
50. *Kochin NYe, Kibel IA, Roze NV* (1963) Theoretical hydromechanics (in Russian). Fizmatlit, Moscow, parts 1, 2
51. *Kozlov VyE, Lebedev AB, Lyubimov DA, Sekundov AN* (2004) Some features of turbulent flow in the leading edge wake (in Russian). Mech Fluid Gas № 1: 78 – 86
52. *Küchemann D* (1978) The aerodynamics design of aircraft, Pergamon, Oxford
53. *Lamb H* (1932) Hydrodynamics, Cambridge University Press, New York
54. *Moin P* (1997) Progress in large eddy simulation of turbulent flows. AIAA, Paper 97-0749
55. *Monin AS, Yaglom AM* (1965) Statistical hydromechanics. Mechanics of turbulence (in Russian). Nauka, Moscow
56. *Nallasamy M, Hussain AKMF* (1985) Numerical study of the phenomenon of turbulence suppression in plane shear layer. In: Turbulent Shear Flows 4, Springer, 8.6 – 8.11
57. *Neimotin LI* (1972) On turbulent mixing of two boundless flows of an ideal fluid (in Russian). In: Applied and theoretical physics, Alma-Ata, Issue 4, 284 – 289
58. *Obukhov AM* (1946) Turbulence in a temperature-nonuniform atmosphere (in Russian). In: Proc of Institute of Theoretical Geophysics of Academy of Sciences USSR № 1: 95 – 115
59. *Ostrovsky GM* (2000) Applied mechanics of inhomogeneous media (in Russian). Nauka, SPb, 360 pp.
60. *Pakin AN* (1997) Application of modified q-ω turbulent model to simulation of dimensional gas motion. Trudy TsAGI № 2627: 79 – 92
61. *Paoli R, Lapote F, Cnenot B* (2003) Dynamics and mixing in a jet/wake vortex interactions. Phys Fluids 15, № 7: 1843 – 1860
62. *Quinn WR* (1992) Turbulent free flows issuing from sharp-edged rectangular slots: The influence of slot aspect ratio. Exper Therm Fluid Sci 5, 203
63. *Sarpkaya T* (2000) New model of vortex decay in the atmosphere. J Aircraft 37, № 1: 53 – 61

64. *Shen S, Ding F, Han J, Lin Y-L, Arya CP, Proctor FH* (1999) Numerical modeling studies of wake vortex real case simulation. AIAA J 99-0755, 16 pp.

65. *Shur M., Strelets M, Travin A, Spalart PR* (1998) Two numerical studies of trailing vortices. AIAA-98-595, 25 pp.

66. *Smagorinsky J* (1963) General circulation experiments with the primitive equations. Monthly Weather Review, 91

67. *Smirnykh EA* (1989) A numerical study with the discrete vortex method of separated flow behind a spoiler (in Russian). Proc TsAGI № 2420: 78 – 86

68. *Spalart FR, Strelets MKh, Travin AK, Shur MP* (2001) Simulation of the turbulent vortex wake behind a high-lift wing (in Russian). Mech Fluid Gas № 5: 64 – 72

69. *Spalart FR, Strelets MKh, Travin AK, Shur MP* (2001) Simulation of interaction between a vortex pair with the ground surface (in Russian). Mech Fluid Gas № 6: 52 – 63

70. *Vlasov EV, Ginevsky AS, Karavosov RK, Frankfurt MO* (1982) Near-wall pressure fluctuations in the separation zone behind 2D obstacles (in Russian). Proc TsAGI № 2137, Ch. 2

71. *Vyshinsky VV,* Physico-mathematical model of aircraft vortex wake in turbulent atmosphere. Doctor thesis. Military air engineering academy, Moscow, 2004

72. *Vyshinsky VV*, Sudakov GG (2006) Vortex wake of an aircraft in a turbulent atmosphere (in Russian). Proc TsAGI № 2667: 155 pp.

73. *Widnall SE* (1975) The structure and dynamics of vortex filaments. Ann Rev Fluid Mech 7, 141 – 166

74. *Winckelmans G, Bricteux L, Cocle R, Duponcheel M, Georges L* (2007) Assessment of multiscale models for LES: spectral behaviour in very high Reynolds number turbulence and cases with aircraft wakes vortices. Proc 5th International Symposium on Turbulence and Shear Flow Phenomena (TSFP-5), 27–29 August 2007, Garching, Germany, vol. 1, pp.. 327-331

75. *Winckelmans G, Cocle R, Dufresne L, Capart R*(2005) Vortex methods and their application to trailing wake vortex simulations. Comptes Rendus Physique (Special issue on aircraft trailing vortices) 6, № 4/5: 467 – 486

76. *Winckelmans G, Cocle R, Dufresne L, Capart R, Bricteux L, Daeninck G, Lonfils T, Duponcheel M, Desenfans O, Georges L* (2006) Direct numerical simulation and large-eddy simulation of wake vortices: going from laboratory conditions to flight conditions European Conference on Computational Fluid Dynamics (ECCOMAS CFD 2006), 5–8 September 2006, Egmond aan Zee, The Netherlands

77. *Winckelmans G, Cottin C, Daeninck G, Leweke T* (2007) Experimental and numerical study of counter-rotating vortex pair dynamics in ground effect. 18e Congrès Français de Mécanique, 27–31 August 2007, Grenoble, France, paper CFM2007–1131

78. *Zamyatin AN, Zavershnev YuA* (1984) Filed studies of the structure and evoluition of the vortex wake of a medium-haul aircraft in the atmospheric surface layer (in Russian). Collection of scientific and technical papers, Flight Research Institute (LII) № 163

79. *Zhelannikov AI* (2002) Vortical situation forecast system (in Russian). Bulletin of the Aviation and Aerostation Academy № 2: 8 – 14

80. *Zhelannikov AI* (2003) Vortex hazard (in Russian). Bulletin of the Aviation and Aerostation Academy № 2: 17 – 21

81. *Zhelannikov AI* (2005) The discrete vortex method and singular integral equations in aerodynamics (in Russian). Integral Equations and Modeling of Applied Problems (IEMAP-2005). Chisinau, Republic of Moldova, 68 – 75

82. *Zhelannikov AI* (2006) An invisible hazard (in Russian). Polyot № 3: 17 – 20

83. *Zhelannikov AI* (2007) A simplified methodology for computing the far vortex wake behind trunk-route aircraft (in Russian). Scientific Bulletin of Moscow State Technical University of Civil Aviation, Aeromechanics and Strength № 111, 16 – 21

84. *Zhelannikov AI, Gulyaev VV, Moroshkin DV, Ushakov SA* (2007) A mathematical model of the far vortex wake behind propeller-driven aircraft (in Russian). Scientific Bulletin of Moscow State Technical University of Civil Aviation, Aeromechanics and Strength № 111: 16 – 27

85. *Zhelannikov AI, Nekrakha SI, Eremenko SM* (2005) A methodology and results of computation of the aerodynamical characteristics of an aircraft with account for terrain-induced vortices (in Russian). In: Methods of discrete singularities in problems of mathematical physics. Int Symp Proc, Issue 4, Oryol, 56 – 62

86. *Zhelannikov AI, Nekrakha SI, Eremenko SM* (2006) A methodology of computation of the aerodynamic characteristics of an aircraft with account for terrain-induced vortices (in Russian). Scientific Bulletin of Moscow State Technical University of Civil Aviation, Aeromechanics and Strength № 97: 6 – 12

87. *Zhelannikov AI, Vyshinsky VV* (2005) Investigation of the characteristics of the trailing wake behind the An-124 aircraft (in Russian). Scientific Bulletin of Moscow State Technical University of Civil Aviation, Aeromechanics and Strength № 81: 19 – 24

88. *Zheng ZC, Lim SH* (2000) Validation and operation of a wake vortex/shear interaction model. J Aircraft 37, № 6, 1037 – 1078

89. *Znamenskaya* IL (1996) Requirements for the size of particles used to visualize the vortex wake behind an aircraft (in Russian). Proc TsAGI №2622: 146 – 150

GPSR Compliance

The European Union's (EU) General Product Safety Regulation (GPSR) is a set of rules that requires consumer products to be safe and our obligations to ensure this.

If you have any concerns about our products, you can contact us on ProductSafety@springernature.com

In case Publisher is established outside the EU, the EU authorized representative is:

Springer Nature Customer Service Center GmbH
Europaplatz 3
69115 Heidelberg, Germany

Batch number: 10704713

Printed by Printforce, the Netherlands